# EDGED WEAPONS OF
# HITLER'S GERMANY

# EDGED WEAPONS OF HITLER'S GERMANY

## Robin Lumsden

MBI Publishing Company

This edition first published in 2001 by MBI Publishing
Company, 729 Prospect Avenue, PO Box 1, Osceola,
WI 54020-0001 USA

© 2001 Robin Lumsden

Previously published by Airlife Publishing Ltd,
Shrewsbury, England.

The information in this book is true and complete to the best
of our knowledge. All recommendations are made without any
guarantee on the part of the author or publisher, who disclaim
any liability incurred in connection with the use of this data or
specific details.

We recognize that some words, model names and designations,
for example, mentioned herein are the property of the
trademark holder. We use them for identification purposes only.
This is not an official publication.

MBI Publishing Company books are also available at discounts
in bulk quantity for industrial or sales-promotional use. For
details write to Special Sales Manager at Motorbooks
International Wholesalers & Distributors, 729 Prospect Avenue,
PO Box 1, Osceola, WI 54020-0001 USA.

Library of Congress Cataloging-in-Publication Data available.

ISBN 0-7603-1131-5

Printed in England.

# Picture/Illustration Credits

| Source | Picture/Illustration No. |
|---|---|
| Author: | 1, 2, 3, 4, 5, 12, 14, 23, 37, 38, 41, 103, 104, 105, 106, 112, 140, 142, 145, 161, 165, 174, 182, 183, 184, 185, 186, 187, 188 |
| J. Angolia: | 115, 126, 137, 138 |
| B. Curtis via T. Johnson: | 139 |
| G. David via T. Johnson: | 123 |
| T. Flesher via T. Johnson: | 19 |
| G. Hughes: | 160 |
| Imperial War Museum: | 53 , 124, 141 |
| T. Johnson: | Covers; 6, 7, 8, 9, 10, 11, 13, 15, 16, 17, 18, 20, 21, 22, 25, 27, 28, 29, 30, 31, 32, 33, 34, 35, 36, 39, 42, 43, 44, 45, 46, 47, 48, 49, 50, 51, 54, 55, 56, 57, 58, 59, 60, 62, 63, 64, 65, 66, 67, 68, 69, 71, 73, 74, 75, 76, 77, 78, 79, 80, 81, 82, 83, 84, 85, 86, 87, 88, 89, 90, 91, 92, 93, 94, 95, 96, 98, 99, 100, 101, 102, 107, 108, 109, 110, 113, 114, 116, 117, 118, 119, 120, 121, 122, 125, 127, 128, 129, 130, 131, 132, 133, 134, 135, 136, 143, 144, 146, 148, 149, 150, 151, 152, 153, 154, 155, 156, 157, 158, 162, 163, 164, 166, 167, 168, 169, 170, 171, 172, 173, 175, 176, 177, 178, 179, 180, 181 |
| E. Leland via T. Johnson: | 72 |
| A. Mollo: | 24, 70 |
| T. Pooler via T. Johnson: | 40 |
| R. Rodachy via T. Johnson: | 159 |
| A. Southard via T. Johnson: | 52, 147 |
| US National Archives via T. Johnson: | 26, 61, 97, 111 |

# CONTENTS

# PREFACE

Between 1933 and 1945, the Third Reich created the widest range of official edged weapons ever authorised by a state for its uniformed military, political and civil personnel. The vast majority of dress daggers and swords were designed and produced during the first six years of Hitler's rule, while the Second World War period witnessed a sharp decline in the use of ceremonial pieces and an explosion in the manufacture of combat bayonets and fighting knives. The factories of Solingen ultimately forged over 50 million blades, which circulated throughout all the occupied territories, from the Arctic Circle to the Mediterranean and from the Channel Islands to the Black Sea, influencing the uniform accoutrements of many countries in the process.

This is the first book to cover the subject of Nazi edged weaponry comprehensively in a single volume. Where possible, illustrations of original examples of the items described have been included, and particular thanks in this respect are due to Lt-Col Tom Johnson, US Army (Retired), who kindly provided pictures of many rare daggers and swords from his extensive archives.

Given the uniform-dominated and militaristic nature of Hitler's Germany, it is perhaps not surprising that the superb quality, striking appearance and sheer variety of the daggers, swords, bayonets and knives described in the following chapters have never since been surpassed in the field of edged weapons.

Robin Lumsden
Cairneyhill

1.  *A Solingen swordsmith at work, 1936.*

# INTRODUCTION

In 1935, the German cinema director Leni Riefenstahl produced her now infamous feature-length documentary *Triumph of the Will*, a propaganda epic depicting events at the Nazi Party rally held at Nuremberg between 5 and 10 September the previous year. One of the shortest but most memorable scenes from the film shows a large group of Hitler Youth boys kneeling in a circle and repeatedly prostrating themselves before an *HJ* knife which is stuck into the ground in their midst. In doing so, they are seen to be honouring not only the weapon itself but also their ancestors, the soil into which the blade is embedded and the German blood it will defend in years to come.

This veneration for edged weapons was something deeply embedded in the German psyche. For centuries, members of the Teutonic tribes of pre-Christian Europe had placed a sword beside each new-born son's cradle to provide him with courage and a warlike spirit in defence of his land, and the ceremony of buckling on the sword was performed to transform the Germanic adolescent into a consecrated warrior. During the Middle Ages, swords became symbolic of power, justice and sovereign authority, and it was customary for oaths to be taken while placing hands on a sword blade. Most German edged weapons of this period had dedications or inscriptions known as 'sword blessings' applied to them to remind the bearer of his sworn duties, and they were ultimately laid to rest with him in his grave.

Sword worship enjoyed a spectacular resurgence during the Gothic revival of the nineteenth century, with the rise of nationalism in the progression towards a unified Reich. A tremendous fascination with medieval history and ancient Teutonic legend was fired by the literary works of right-wing authors, poets and composers, particularly the anti-Semite Richard Wagner (1813–83), whose early operatic heroes such as Parsifal and Lohengrin were the epitomes of knightly chivalry, continually battling against the forces of evil. Wagner's last and most magnificent work, *The Ring of the Nibelung*, was set in the murky world of Dark Age fables and was acted out in a land of gods, giants, dragons, supermen and slavish sub-human

**2.** *A dedication being applied to the blade of a presentation sword, using acid-resistant wax. Once this process had been completed, the blade was immersed in a bath of acid which ate into the surface, following the contours of the wax resist, resulting in the desired permanently etched pattern.*

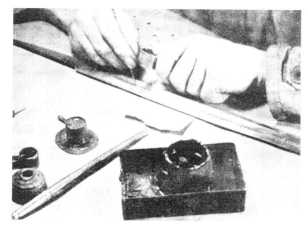

dwarfs, where an enchanted Excalibur-like sword called Nothung bestowed invincibility upon Siegfried, its forger and owner. The new cult of the sword in Germany grew to be almost as strong as the old one in Japan, and it was seized upon by the Nazis in 1933. Hitler quickly resolved to instil a feeling of might and superiority in his uniformed followers by rewarding them with quality edged weapons which, although mass-produced, would hark back to the elite warrior aristocracy of Germany's past. To achieve this aim he turned to the craftsmen of Solingen, the 'City of Blades'.

Located in the Wupper Valley, part of the Ruhr's industrial heartland, Solingen had been famous for its cutlers and swordsmiths since the fifteenth century. The local proof mark of the running wolf, and the blade epithet *Me Fecit Solingen* ('Solingen made me') became highly regarded throughout the whole of the medieval world. Families like the Weyersbergs and Kirschbaums initially controlled powerful cartels which severely restricted the city's edged weapon production with a view to maximising prices. However, the many European wars fought between 1500 and 1700 created a considerable demand for quality Solingen blades and an even bigger clamour for skilled swordsmiths, and the grip of the cartels was gradually relaxed so as to stem the flood tide of trained German cutlers moving to Milan, Toledo and further afield for work. So it was that the golden age of Solingen came about in the eighteenth century, when the city boasted an unprecedented concentration of swordmakers of enviable reputation.

The development of the steam engine opened up new challenges and opportunities for Solingen's craftsmen, with the arduous labour of forging a blade by hand over an anvil being steadily replaced by the use of mechanically driven hammers which could relentlessly pound the glowing billets of metal into shape. Even more dramatic was the evolution of mass production, as blades began to be drop-forged in dies, then drawn and shaped mechanically, with only the finishing work being left to the hand labourer. Solingen's many small and independent sword-making enterprises soon had to group together into co-operative ventures to

**3.** *A cutler and his apprentices discuss the finer points of a Navy sabre.*

handle the ever larger orders being placed by both domestic and foreign governments, particularly as state contracts demanded new and uniform standards with which the traditional hand-forging bladesmiths could not hope to comply. All this added impetus to the wholesale gathering together of craft workers under corporate roofs and the amalgamation of old family names, exemplified by the foundation of the Weyersberg-Kirschbaum company in 1883. With the advent of the First World War, unprecedented pressures were placed on the blade firms of Solingen to produce standardised weapons in vast quantities. This resulted in the development of a process of pre-tempering and rolling steel into sheets from which multiple blades were cut, ground and finished automatically. Tens of millions of swords, bayonets and fighting knives were manufactured in this way, keeping the city's forges fully occupied day and night. However, at the end of 1918 blade production came to a sudden and total halt as the victorious Allies determined to destroy the former Kaiser's war machinery. Solingen, like the rest of Germany, was thrown into a condition of economic depression and had to earn its living during the 1920s by producing forks, spoons, needles, razors and other basic articles for household consumption.

The election of Adolf Hitler as Chancellor on 30 January 1933 heralded a bright new era of blade manufacture in Solingen. Mussolini's Fascist paramilitary formations had started wearing impressive dress daggers in the early 1920s, and this feature of their uniform did not escape the envious attention of the Nazis. As early as 1928, Hitler directed that all members of his new youth organisation were obliged to kit themselves out with 'hiking knives' of a then unspecified description. This doubtless came as a welcome surprise to the businessmen of Solingen, who proceeded to meet much of the short-term Hitler Youth demand simply by adding swastika badges to surplus stocks of assorted short-bladed fighting knives left over from the First World War. It quickly became evident that within the expanding Byzantine structures of the National Socialist German Workers' Party (*NSDAP*) there existed other vast uniformed organisations which likewise sought the adornment of dress daggers and swords. The order for Hitler Youth knives was soon followed by others placed on behalf of political bodies such as the brown-shirted *Sturmabteilung* (*SA*), with membership numbers running into the millions, and this was only the tip of the iceberg. When Germany finally shook off the shackles of Versailles and

**4.** *Main entrance block to the Eickhorn factory in 1940.*

embarked upon a war footing from 1935, the metalware industries of the Ruhr once again had a fulltime job as their presses resounded with the production of blades and equipment for the new *Wehrmacht*.

Solingen sustained considerable bomb damage during 1944, by which time its production of dress items had all but ceased. The following year, after disarming the *Wehrmacht*, the Allied Military Government in Germany decreed that all weapons still in the hands of the civilian population should also be surrendered. Most people complied, and this yielded vast quantities of ornate edged weapons which soon lay in huge piles on street corners all over the country. Dedicated teams of Allied soldiers were thereafter engaged in the indiscriminate destruction of this material, two favoured methods being to crush blades under the wheels of vehicles and to load railway goods carriages to capacity and then run them into local lakes. Consequently, only a very small proportion of Solingen's vast Nazi output survived after 1945, in the form of items not surrendered and those selected at random from the destruction process to be taken home by Allied troops as souvenirs.

The edged weaponry of Hitler's Germany comprised an unprecedented selection of military, political and civil daggers, swords, bayonets and knives, complemented by their own distinctive hangers, knots and other accoutrements. All were carefully designed so as to be both modern and yet firmly rooted in the nation's past, thereby typifying that mix of revolutionary and traditional values which was so characteristic of the Third Reich. The following chapters describe these weapons in turn, along with the organisations to which they related.

# PART ONE
# DAGGERS

Prior to 1933, only a few regulation German daggers existed for wear by naval officers, selected higher ranks of the Fire Brigade, Forestry Service officials, members of hunting and shooting associations and boys from various youth movements. By 1941, over twenty uniformed organisations had their own patterns of dagger, issued in more than fifty different styles for different categories of wearer. Some were made in vast numbers, while others were produced in extremely limited quantities or even on a one-off basis for special presentation purposes.

The inspirations for the designs of these impressive new edged weapons were many and varied. The bulky Labour Service hewer, for example, was based on an old Germanic hatchet, while the early Air Force dagger had a slim and distinctly medieval look. The clean neo-classical lines of the Army officers' version contrasted sharply with the Renaissance curves of *SA* and *Schutzstaffeln* (*SS*) weapons, whereas the daggers worn by diplomats, government officials and staff of the Air Raid Protection League were typically 1930s art deco in design.

Whatever its origin or pattern, however, virtually every Nazi dagger bore the swastika as undisputed emblem of the new Reich.

## ARMY

The humiliating Versailles Treaty of June 1919 limited Germany's standing army to 100,000 men, of whom only 4,000 were to be officers. Poison gas, tanks and heavy weapons were prohibited, military

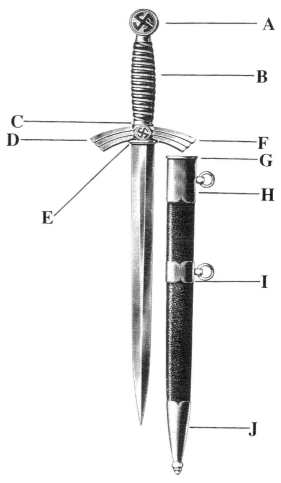

**5**. *Dagger nomenclature (*Luftwaffe *1935 dagger as example): A - Pommel; B - Grip; C - Ferrule/bolster; D - Quillon; E - Quillon block; F - Crossguard; G - Throat; H - Locket; I - Central mount; J - Chape*

academies closed and the old Imperial General Staff system abolished. However, when the new national army, or *Reichsheer*, was formally inaugurated four months later, its much reduced size actually worked to its advantage, ensuring that only the very best and most competent professional officers and NCOs were retained in service.

During the period 1919–33, promotion in the Army was extremely slow and some officers were even obliged to accept reductions in rank in order to maintain their employment. Throughout his struggle for power, Hitler consistently wooed the *Reichsheer* leadership by promising military expansion, and so by supporting him the officer corps naturally hoped to regain their former status and influence. General Werner von Blomberg, a Nazi sympathiser, became War Minister in Hitler's first cabinet and he in turn ensured that all key Army posts went to supporters of the new regime. In May 1934 the *NSDAP* eagle and swastika insignia, or *Hoheitsabzeichen*, became a feature of Army uniform, and the following August the military took an oath of loyalty to Hitler personally, rather than to the state. The Army was duly rewarded for its support by an increase in size from seven to twenty-one divisions. A much bigger expansion was heralded at the beginning of 1935, with the enactment of the Law for the Reconstruction of the National Defence Forces. This legislation introduced conscription for all men of military age from 16 March that year, and necessitated a rapid development of the armaments industry to equip the greatly enlarged armed forces, or *Wehrmacht*. The compulsory period of military service was initially set at one year, but in August 1936 it was increased to two years. By that time the Army had grown to thirty-six divisions, with the latest tanks and artillery, and had been retitled *Heer*.

While strengthening the manpower and hardware of the Army as a whole, Hitler shrewdly took great pains to pander specifically to the wounded vanity of the officer class by giving them smarter uniforms and insignia, a whole range of badges and awards, and other incentives designed to restore their confidence and retain their loyalty. Among these trappings was a new dress dagger, which

ultimately saw wear by some 300,000 commissioned personnel of all ranks.

## THE ARMY 1935 DAGGER

Approved by the *Heeresleitung* on 4 May 1935, the Army dagger was designed by Paul Casberg, a 52-year-old military artist who worked closely with the Carl Eickhorn firm and created many Third Reich edged weapons, flags and badges during the course of his career. Casberg submitted a number of drawings showing his proposals, some of which were very ornate. The one selected by Hitler was clean in line and pure in form, with a simple oakleaf pommel and eagle crossguard. Production by many different companies began almost immediately, and an order published on 18 June laid down the correct manner of wear for the dagger and the style of its accompanying portepée knot and hanging straps. Large stocks of the weapon and accoutrements were quickly distributed to military outfitters throughout Germany. The new dagger replaced the sabre and pistol with walking-out dress, and it had to be purchased privately by all

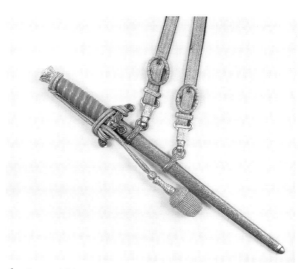

**6.** *Army 1935 dagger. This example features the yellow grip favoured by cavalry officers, and ornate oakleaf-embossed hanger fittings which were available at extra cost. The aluminium portepée knot is tied in the manner prescribed on 4 May 1942.*

zinc-based alloy. The regulation plastic grip was cream-white in colour, to simulate ivory, but large numbers of yellow versions were also produced for cavalry and signals officers, to match the *Waffenfarbe* piping on their uniforms. A few grips were also made in orange-red and black, for artillery and engineer officers respectively.

While the basic dagger cost 11.50 Reichsmarks, there were many more costly non-standard variations available, depending upon the price the buyer was prepared to pay. For an extra 10.20 Reichsmarks he could have a real ivory grip, while etched blade inscriptions were available from 0.60

**7.** *An Army Leutnant, a veteran of the first winter campaign in Russia, posing proudly with his dagger during the autumn of 1942.*

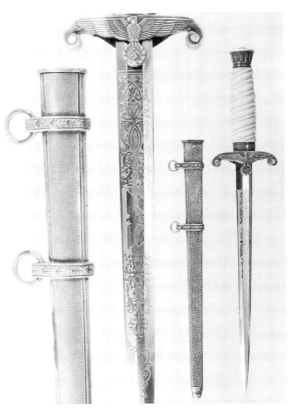

**8.** *Army officers could opt to have their dagger blades etched with any one of a range of appropriate designs. The pattern illustrated here, produced by Alcoso, is typical and depicts floral decoration around an eagle and swastika. It is shown alongside a matching miniature dagger, of the sort used by Alcoso's travelling salesmen as samples.*

army officers, including those holding reserve or specialist commissions. From 19 January 1937, it could also be bought and worn by officer cadets and by NCOs in the latter stages of officer training.

The Army dagger measured 40 cm in total length, with a 26 cm plain steel blade, and its silvered fittings were artificially oxidised to give a tarnished appearance. Early examples were plated, while later pieces were produced from a lacquered dull grey,

**9.** *The original owner of this Army dagger, manufactured by Eickhorn between 1938 and 1941, had an elaborate monogram of his initials, 'A.S.', engraved on the reverse of the crossguard.*

On 27 May 1943, Armaments Minister Albert Speer ordered an end to the manufacture of Army daggers for the duration of the war for economic reasons. Officers commissioned after that date reverted to the wearing of holstered pistols if they could not easily acquire a pre-stocked or second-hand dagger. Further use of the dagger was totally prohibited on 23 December 1944, when all officers were instructed to carry pistols with walking-out dress, in keeping with the critical war situation at that time.

### THE ARMY 1940 HONOUR DAGGER

On 28 December 1940 a unique Army Honour Dagger, or *Ehrendolch des Heeres*, was presented

**10.** *The Army Honour Dagger presented to SA-Stabschef Viktor Lutze by his father-in-law, Generalfeldmarschall von Brauchitsch, on 28 December 1940.*

Reichsmarks per letter and engraved monograms or initials from 1 Reichsmark. Those with more expensive tastes could obtain genuine Damascus steel blades for around 70 Reichsmarks.

The hanging straps for the Army dagger were made from field-grey velvet faced with aluminium bullion wire, and bore silvered or white metal fittings. Gold plating was used by general officers from October 1942. Once again, several finely decorated hanger variants existed, with prices to match. The accompanying portepée knot comprised a 42 cm strap and 'acorn', produced from aluminium wire until 17 January 1942 and thereafter from a cheaper grey-white rayon called Celleon. On 4 May 1942 a revised manner of wear was prescribed for the knot to reduce the deterioration caused by constant rubbing against the dagger crossguard. The production of dress knots ceased in April 1943.

by *Generalfeldmarschall* von Brauchitsch to *SA-Stabschef* Viktor Lutze on the latter's fiftieth birthday, in recognition of the contribution made by the *SA* in the field of pre-military training. The hand-finished dagger measured 6 cm longer than the standard version and featured hallmarked silver fittings, a massive eagle pommel, a white ivory grip wrapped with silver wire, and a Damascus steel blade with the raised gilt inscription *Das Deutsche Heer dem Stabschef Lutze, 28.12.40 – Treue um Treue*. Its ornate hanging straps were heavily embroidered with oakleaves and bore a large *Wehrmacht* eagle in silver bullion on a scarlet cloth central shield.

It is interesting to note that while Lutze possessed several ornate edged weapons given to him by different military and paramilitary organisations, including the Honour Dagger of his own *SA*, it was the *Ehrendolch des Heeres* which was selected to lie on top of his coffin on the occasion of his state funeral in May 1943.

Three similar Army Honour Daggers, with different-coloured fittings and suitable blade dedications, are known to have been gifted to Adolf Hitler, *NSKK-Korpsführer* Adolf Hühnlein and *Reichsarbeitsführer* Konstantin Hierl.

# NAVY

The Versailles settlement imposed strict conditions upon the *Kaiserlichmarine*, and directed that the vast majority of its vessels be confiscated by the victorious Allied powers. This prospect so outraged the Naval High Command that the admirals ordered the entire imperial fleet to be scuttled whilst anchored at Scapa Flow in the Orkneys. As a result, the infant *Reichsmarine* of the Weimar Republic was left with a clean slate on which to redesign its warships within the 10,000 tonne gross weight limitation laid down at Versailles. Through skilful design work on the part of Blohm und Voss and other German shipbuilders, pocket battleships such as the *Graf Spee* were conceived, which retained their heavy-calibre deck guns but kept within the weight limit by having drastically reduced levels of armour plating. Consequently, they could easily exceed the speed

capabilities of any other warships in the world carrying comparable armaments. In 1935, Hitler successfully negotiated an Anglo-German Naval Agreement effectively lifting the Versailles restrictions and allowing the construction of revolutionary 46,000 tonne capital ships, including the *Bismarck* and the *Tirpitz*. A U-boat building programme was also inaugurated and the new *Kriegsmarine* was on the ascendant. However, the naval officer corps retained its traditional and conservative outlook, which undoubtedly accounted for the fact that the Navy was the last of the three *Wehrmacht* services to incorporate Nazi symbolism into its edged weapons.

## THE NAVY 1938 DAGGER

The Prussian and German navies had a long history of wearing dress daggers dating back to 1849, and during the nineteenth and early twentieth centuries their basic appearance changed very little. Initially, dirks had been worn only by cadets, but in 1901 their use was extended to officers. On 28 November 1919, a black grip and a black scabbard with a single carrying ring were introduced, and the imperial crown pommel was superseded by one featuring a flaming ball. The black scabbard was discarded on 5 April 1921 and the old gilded scabbard and two carrying rings were reintroduced, while on 5 June 1929 the black grip was replaced by a traditional white version. However, the flaming ball pommel continued to be worn five years into the Third Reich.

On 20 April 1938, to coincide with Hitler's birthday, the Naval High Command introduced the definitive Nazi Navy dagger, which replaced the flaming ball with an eagle and swastika pommel. The regulation weapon was made from brass, heavily gilded and lacquered, with its scabbard either hammered or engraved with lightning bolts. The crossbands for the hanger rings could feature an oakleaf or twisted rope design, while the 25 cm steel blade was normally etched with ships, anchors or similar motifs. The dagger was locked into its scabbard by means of a small brass stud on the reverse of the fouled anchor and acanthus-leaf crossguard. The grip was typically in white

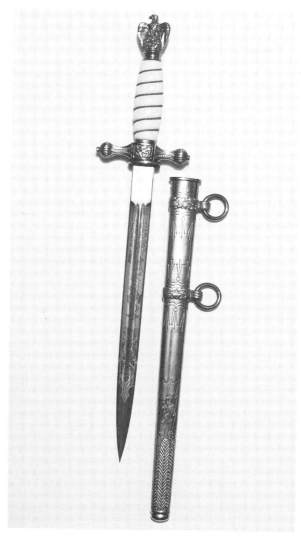

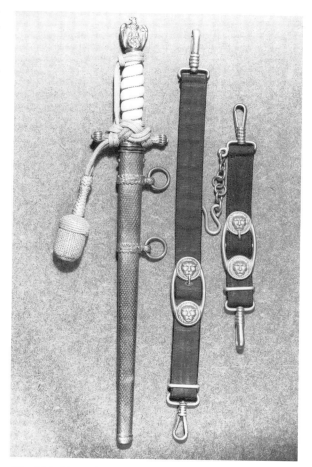

**12.** *This Navy dagger by WKC has the scarcer hammered scabbard. The portepée knot is tied in the regulation way, and the hanging straps display their traditional lion-mask buckles.*

**11.** *Standard Navy 1938 dagger by F.W. Höller, with engraved scabbard.*

celluloid, bound with twisted double-strand brass wire, although yellow versions were also produced to complement the gold insignia of the naval uniform.

An array of variant Navy daggers was manufactured for private purchase, and the vastness of the selection available was compounded by the updating or altering of weapons over the years as they passed from father to son, or as individual naval officers served under the imperial, republican and Nazi regimes. For example, those who had fought in the First World War were permitted to wear crown pommels during the Third Reich, and the crowns were reproduced after 1938 for that reason. The basic dagger normally cost around 18 Reichsmarks with an engraved scabbard, or 19.05 Reichsmarks with a hammered scabbard, while a real ivory grip added 8 Reichsmarks to the price.

Damascus steel blades ranged from 18 Reichsmarks for plainer versions to 70 Reichsmarks for the more decorative 'Turkish' patterned types.

The hangers for the Navy dagger took the form of two separate velvet-backed black moiré silk straps with lion-mask buckles and snap fasteners at each end. A small chain was attached to hold the dagger in a vertical position if required. The brass or alloy fittings were gilded for line officers and silvered for administrative personnel. A 42 cm aluminium or gold wire portepée knot was authorised for use by all officers and by cadets who had passed their qualifying examinations. It was tied in a unique style, known erroneously as the 'reef knot'.

**13.** *Naval* Kapitänleutnant *Armin Zimmermann wearing his dagger at the time of the surrender of the German forces on the Channel Islands, May 1945. Its scabbard clearly exhibits the typical dent caused when swinging daggers were caught in the sliding doors of railway carriages.*

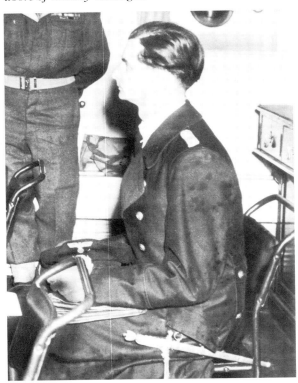

**14.** *Navy dagger blade by WKC, etched with a fouled anchor, a sailing ship and acanthus leaves.*

Albert Speer's War Economy Order of 27 May 1943 made no reference to halting the production of Navy daggers. However, on 5 November that year existing stocks of daggers held at the various naval supply depots throughout the Reich were recalled and centralised at Kiel, and from 25 February 1944 the wearing of the weapon with the field-grey marine uniform was discontinued. On 23 December 1944, in common with those of the other *Wehrmacht* services, Navy daggers ceased to be used, being replaced by pistols.

Of over 100 firms which were engaged in the manufacture of Nazi edged weapons, only seventeen produced the Navy dagger: Alcoso, Clemen & Jung, Eickhorn, Höller, Horster, Robert Klaas, Krebs, Lauterjung, Lüneschloss, Pack, Plumacher, Puma, Paul Seilheimer, Max Weyersberg, Paul Weyersberg, WKC and Winger. In their sales catalogues of 1938–40, the Pack and Puma companies advertised an alternative model of Navy dagger bearing acanthus leaves on the pommel and an eagle and swastika on the crossguard. This design was never approved by the *Kriegsmarine* and does not appear to have been popular with potential buyers, as no photographs of it being worn have ever come to light. Indeed, it may have existed only as an artist's drawing or in a prototype form.

## THE NAVY 1938 HONOUR DAGGER

On 20 April 1938, naval Commander-in-Chief *Grossadmiral* Erich Raeder created the *Kriegsmarine-Ehrendolch* to recognise exceptional merit on the part of his officers. Designed by Paul Casberg, the Honour Dagger was produced by Carl Eickhorn and had the pommel swastika inset with seventeen diamonds. The grip was bound with gilt oakleaves, the quillon block was larger, and the scabbard was decorated with oakleaves and acorns. The rose-pattern Damascus steel blade bore a suitable presentation dedication.

The first Honour Dagger was presented by Raeder on 31 December 1939 to *Vizeadmiral* Conrad Albrecht, who had successfully directed *Kriegsmarine* operations against the Polish fleet. By the end of 1942, a further five examples had

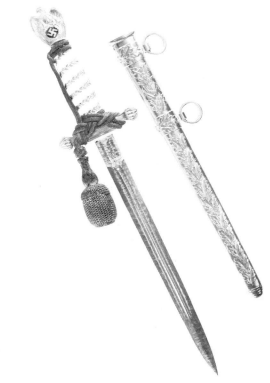

**15**. *The Navy 1938 Honour Dagger presented by* Grossadmiral *Erich Raeder to U-boat ace* Fregattenkapitän *Reinhard Suhren on 2 September 1942, the day after the latter won the Swords to the Knight's Cross of the Iron Cross. Of particular note are the diamond-encrusted pommel swastika, the oakleaf banding to the ivory grips, the ornate scabbard and the Damascus steel blade. The blade ricasso bears the unique dedication* Dem U-Bootsieger – Raeder – 2.9.1942.

been conferred, upon Günther Prien of U-47, Erich Topp of U-552, Reinhard Suhren of U-564, Naval Armaments Quartermaster Carl Witzell and Admiral Alfred Saalwächter. Raeder's successor, Karl Dönitz, gifted a seventh on 9 May 1944 to Albrecht Brandi of U-967. No further awards are known to have been made.

Exact criteria for presentation of the Honour Dagger were never published, and distribution appears to have been at the personal discretion of Raeder and Dönitz. However, it is noteworthy that only serves to illustrate the desirability of such items for collectors, and the depths to which some will sink to obtain them.

# AIR FORCE

During the First World War, the German air service, or *Luftstreitkräfte*, was merely an auxiliary formation of the Army, and it was completely disbanded by the Allies in 1919. Hitler had been greatly impressed by the destructive power of aircraft during the conflict, and on becoming Chancellor in 1933 one of his primary concerns was the formation of the German Air Sports Association, a façade behind which civilian pilots were instructed in military aviation techniques. Its members formed the cadre of the new *Luftwaffe*, which was established on 26 February 1935 as an independent branch of the *Wehrmacht*. Lacking tradition and owing its existence almost entirely to Hitler and Göring, the *Luftwaffe* developed into the most fervently National Socialist of all the German fighting services, excepting the *Waffen-SS*.

### THE *LUFTWAFFE* 1935 DAGGER

In March 1935, a *Fliegerdolch*, or airman's dagger, was introduced for wear by all *Luftwaffe* officers and senior NCOs. Based on a similar edged weapon already worn by personnel of the German Air Sports Association, the 48 cm *Fliegerdolch* was medieval in appearance, with its wooden grip and scabbard covered in dark blue Morocco leather. The grooved handle was wrapped diagonally with a triple silver-coloured wire, while the circular pommel featured an inlaid brass sunwheel swastika, as did the down-swept wing crossguard. All metal fittings were initially in nickel silver, then polished aluminium from 1936. Priced at 17 Reichsmarks, the dagger was suspended from a plain double chain hanger comprising nickel silver or aluminium rings. A 23 cm silver-coloured portepée knot was authorised for use by flying personnel. Following the introduction of the *Luftwaffe* officer's dagger, the *Fliegerdolch* continued to be worn only by junior NCOs and by other ranks who were entitled

**16**. *Army* Generaloberst *Eduard Dietl wearing the singular Navy Honour Dagger presented in recognition of his leadership of combined operations during the Battle of Narvik in 1940. A miniature Destroyer War Badge is mounted between the scabbard bands, with an edelweiss emblem under the lower band. The weapon is suspended from standard Army hangers.*

Prien was given the dagger upon his receipt of the oakleaves to the Knight's Cross of the Iron Cross, while Topp, Suhren and Brandi got theirs when they won the swords to the Knight's Cross. As the war dragged on, and the holding of senior military decorations became less uncommon, so bestowals of the Honour Dagger tailed off.

In August 1999, 85-year-old Erich Topp welcomed a stranger into his home at Remagen to view his war souvenirs. The man repaid his trust by stealing Topp's Honour Dagger, Knight's Cross with oakleaves and swords, and citations. This sad tale

**17.** *A Luftwaffe 1935 dagger. This early production piece has nickel silver fittings and matching chain hanger, and is decorated with a DLV-style 42 cm portepée knot instead of the 23 cm Air Force version.*

**18.** This Luftwaffe Gefreiter, *an air gunner and NCO candidate, wears the Air Force 1935 dagger on the occasion of his wedding, 16 September 1943.*

to one of the various aircrew qualification badges, such as the Pilot Badge or the Radio Operator Badge. In common with other military daggers, use of the *Fliegerdolch* was finally forbidden on 23 December 1944.

**19.** *Unit property stamp 'LUFT/101K' on the locket of a* Luftwaffe *1935 dagger.*

**20**. *Hermann Göring personally influenced the design of the pseudo-medieval* Fliegerdolch, *and wore it proudly by his side at many social events during the mid-1930s.*

**21**. *A Luftwaffe 1937 dagger with a yellow grip.*

## THE *LUFTWAFFE* 1937 DAGGER

On 1 October 1937, Göring autho-rised a new *Luftwaffe Offizier-dolch*, to be worn by officers instead of the *Fliegerdolch*. Its use was extended to senior NCOs on 12 March 1940. The style of the *Offizierdolch* was a complete departure from that of its prede-cessor, and the weapon was more akin to the daggers carried by Army and Navy officers. It measured 41 cm in length, with a plain steel blade, and had cast aluminium or alloy hilt fittings and a grey steel scabbard. The regulation plastic grip was white in colour, wrapped with a twisted silver wire, although yellow versions were also made to match the *Waffenfarbe* piping

**22**. *As a senior NCO, this* Feldwebel *radio operator was entitled to wear the* Luftwaffe Offizierdolch.

worn by flying personnel and paratroops. The pom-mel bore a prominent gilded swastika, while the crossguard comprised a massive flying eagle repre-sentative of the Air Force. The basic *Offizierdolch* cost 13.50 Reichsmarks, but as with Army and Navy daggers there were additional extras such as ivory grips and Damascus blades which buyers could obtain at a higher price.

The hanging straps for the officer's dagger were made from *Luftwaffe* blue-grey cloth faced with sil-ver woven edges, sewn to a velvet backing. Their rectangular hanger buckles bore an oakleaf design and were in grey metal, or, after 31 May 1942, gild-ed metal for general officers. The accompanying portepée knot measured 23 cm in length and was identical to that used with the *Fliegerdolch*. Once

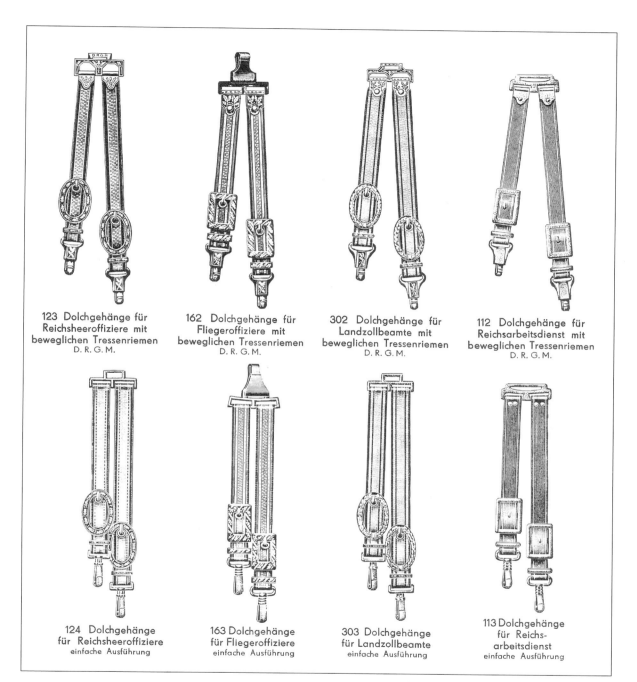

123 Dolchgehänge für
Reichsheeroffiziere mit
beweglichen Tressenriemen
D. R. G. M.

162 Dolchgehänge für
Fliegeroffiziere mit
beweglichen Tressenriemen
D. R. G. M.

302 Dolchgehänge für
Landzollbeamte mit
beweglichen Tressenriemen
D. R. G. M.

112 Dolchgehänge für
Reichsarbeitsdienst mit
beweglichen Tressenriemen
D. R. G. M.

124 Dolchgehänge
für Reichsheeroffiziere
einfache Ausführung

163 Dolchgehänge
für Fliegeroffiziere
einfache Ausführung

303 Dolchgehänge
für Landzollbeamte
einfache Ausführung

113 Dolchgehänge
für Reichs-
arbeitsdienst
einfache Ausführung

**23**. *A selection of hanging straps advertised for sale by the Solingen firm of Röder & Co. From left to right, they relate to: the Army 1935 dagger; the* Luftwaffe *1937 dagger; the Land Customs 1937 dagger; and the RAD leader's 1937 hewer. Those on the top row feature elaborate fittings available at extra cost, while those on the bottom are of more basic manufacture.*

again, further wear of this dagger was prohibited on 23 December 1944.

## THE *REICHSMARSCHALL* DAGGER

On 19 July 1940, Hermann Göring was elevated to the newly created and extraordinary rank of *Reichsmarschall des Grossdeutschen Reiches*, which effectively made him the most senior *Wehrmacht* officer, subordinate only to Hitler. In a nation that was besotted with the wearing of military and paramilitary uniforms of all kinds, Göring stood out from all others in his love of uniform attire in general, and edged weapons in particular. To distinguish his promotion to *Reichsmarschall*, he had a new uniform designed and tailored for himself in dove-grey material, which had no precedent within the German armed forces. Several versions of it were eventually produced, all incorporating the *Luftwaffe* eagle – although his unique rank was a *Wehrmacht* one, not restricted to the Air Force.

To complement Göring's new outfit, students of the Berlin Technical Academy under the direction of Professor Herbert Zeitner produced a gold *Reichsmarschall-Dolch* featuring a white ivory grip and a pommel and crossguard set with a large number of rubies and diamonds. The *Reichsmarschall*'s eagle with crossed batons insignia and representations of the Nazi Grand Cross of the Iron Cross, of which Göring was the sole recipient, were prominent aspects of the design. The accompanying hangers took the form of white velvet straps faced with gold and silver bullion. While never apparently worn, the fabulous *Reichsmarschall* dagger was described in issue no. 22/1940 of the trade magazine *Uniform-Markt* and was displayed at an exhibition of German war booty staged in London in November 1945, after which it disappeared. It has never been seen since.

## SA

The *Sturmabteilung*, or *SA*, was formed by Ernst Röhm and Hans-Ulrich Klintzsch during 1921, as a strong-arm squad to fend off opponents at Nazi

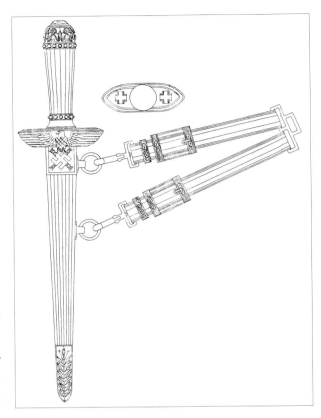

**24.** *The* Reichsmarschall *dagger.*

Party political meetings. Named rather grandiosely after the 'assault detachments' of elite stormtroopers who had fought on the western front during the First World War, the embryonic *SA* was originally confined to Munich. It made its first important foray outside the city on 14–15 October 1922, when it took part in a 'German Day' at Coburg which resulted in a pitched street battle with the Communists who controlled the town. The 800 *SA* men present, almost the entire membership at that time, succeeded in breaking the hold of the Red Front on Coburg, and press coverage of the incident served to make Hitler's name known throughout Germany for the first time.

Röhm soon made himself indispensable to Hitler through his ability to obtain arms and supplies via his military connections. He came to envisage his

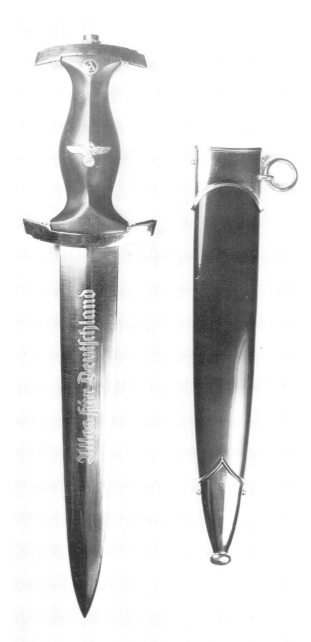

**26**. *An* SA *man from Gruppe 'Hochland' wearing his dagger in the vertical position during a sightseeing trip to the Bavarian Alps, 1935.*

**25**. *An early example of the* SA *1933 dagger, characterised by its fine-quality nickel silver fittings and anodised scabbard. Note how the etched motto* Alles für Deutschland *is closer to the hilt than to the tip of the blade – a feature common to all originals.*

*SA* as a citizen army which could bolster and ultimately supersede the *Reichsheer,* and it grew rapidly during the 1920s. By 1934 it was forty times the size of the regular Army and included cavalry regiments, naval battalions and air squadrons. However, denied the position of real power to which they felt entitled in the new Nazi state, Röhm and his senior *SA* commanders began to talk of a 'second revolution' which would sweep away the bourgeois in the party and the reactionaries in the Army. Hitler knew he could never achieve supreme authority without the backing of the military and so decided that the *SA*, having served its purpose, would have to be cut down to size. The danger it posed was just too great – not only the threat of a *putsch* but also the ever-present disorder created by the very men who should have been setting an example of good order. Their incessant brawling, drinking and violence, to say nothing of Röhm's notorious homosexual antics, provoked profound discontent among the public. The confidence ordinary Germans had in the new regime

**27.** *This new recruit to the elite* SA-Standarte Feldherrnhalle *wears his dagger suspended from a hanger located beneath the lower left pocket flap of the tunic, 1938.*

was in danger of collapsing altogether. On 28 June 1934, therefore, Hitler took the final decision to eliminate the *SA* leadership. Two days later he personally directed operations at Munich and Bad Wiessee when armed *SS* formations arrested and executed Röhm and sixteen of his highest officers. Over 300 others also lost their lives in this bloody purge, which was to become known as the Night of the Long Knives. The *SA* suffered a loss of power and influence from which it never fully recovered. Röhm's successor, Viktor Lutze, was obsequiously loyal to Hitler and respectful to the generals, while the rank and file of *SA* members were reduced from 4 million to just over 1 million of the better elements, stripped of their arms.

After 1939, the *SA* enjoyed a moderate resurgence, co-ordinating pre- and post-military training programmes in liaison with the *Wehrmacht*. During the latter stages of the Second World War it was also given responsibility for setting up home guard units in the occupied territories. However, the wartime *SA* always remained a very pale shadow of its former self, the gigantic force of no-nonsense street fighters whose brown-shirted members could be seen swaggering about in gangs in every corner of Germany during 1933–4.

As the oldest paramilitary formation of the Nazi Party, the *SA* was the first uniformed organisation to be accorded the honour of a dress dagger following Hitler's accession to power. At that time the *SS* and the National Socialist Motor Corps were part of the larger *SA*, and this was reflected in the common design of the daggers worn by the three groups.

## The *SA* 1933 Dagger

During the summer of 1933, Hitler commissioned Professor Woenne of the Solingen School of Commerce to design an edged weapon for wear by members of the *SA* and *SA* Reserve. Introduced on 15 December that year by order of the acting *SA* Chief of Staff, *Obergruppenführer* von Krausser, the resulting *SA-Dolch* was styled after a sixteenth-century south German hunting knife known as the Holbein, in the collection of Munich City Museum,

28. *This SA 1933 dagger by Carl Wüsthof bears the 'Nm' inspection stamp of* SA-Gruppe *'Nordmark' on the centre of the crossguard reverse.*

which featured a representation of Holbein's painting 'The Dance of Death' on its scabbard. The new dagger was a heavy quality item measuring 37 cm in length. Its curved grip was made from a variety of woods, primarily pear, walnut and maple, and was inlaid with a nickel eagle and swastika and the enamelled runic insignia of the *SA*. The fine steel blade bore the etched dedication *Alles für Deutschland* ('Everything for Germany') in Gothic script, while the seamless sheet metal scabbard was finished in a lacquered red-brown oxide. Both the hilt and scabbard fittings were in nickel silver.

A short, detachable brown leather strap with an oval nickel buckle at one end and a clip at the other was used to suspend the weapon diagonally from the uniform belt. An additional strap could be affixed round the top of the grip, if required, to hold the dagger in a steadier vertical position while marching.

The *SA* dagger, costing 7.30 Reichsmarks, could not initially be bought from retail outlets but had to be ordered and paid for by members through their local units. Pieces so supplied between 1933 and 1935 were subject to quality-control inspection at the headquarters of the appropriate *SA-Gruppe* prior to being delivered to the buyer, and each was stamped on the reverse crossguard with one of the following identifying headquarters abbreviations, indicating that this process had been carried out:

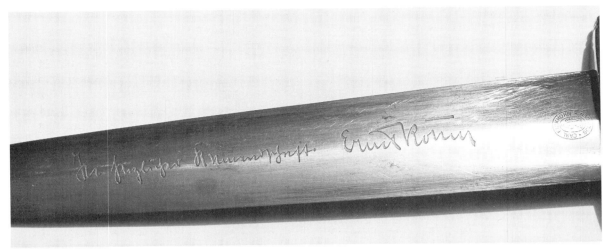

**29**. *One of the few surviving examples of the complete SA 1934 Honour Dagger blade dedication,* In herzlicher kameradschaft, Ernst Röhm.

| INSPECTION STAMP | HQ OF *SA-GRUPPE* |
|---|---|
| B | Berlin-Brandenburg |
| BO | Bayerische Ostmark |
| Fr | Franken |
| Ha | Hansa |
| He | Hessen |
| Ho | Hochland |
| Kp | Kurpfalz |
| Mi | Mitte |
| Nm | Nordmark |
| Nrh | Niederrhein |
| Ns | Niedersachsen |
| No | Nordsee |
| Om | Ostmark |
| P | Pommern |
| S | Schlesien |
| Sa | Sachsen |
| Sw | Südwest |
| Th | Thüringen |
| Wf | Westfalen |
| Wm | Westmark |

From 1936, the production and supply of *SA* daggers came under the control of the *Reichs-zeugmeisterei,* or *RZM,* the Nazi Party contracts office. Daggers could thereafter be bought by individual *SA* men directly from *RZM*-approved shops. Consequently, the use of *Gruppe* inspection stampings ceased and the *RZM* logo and maker's code number etched on the reverse side of the blade in place of the usual company trademark was taken as evidence that the dagger met the rigorous quality standards which were laid down nationally. These post-1936 daggers generally featured lighter and less expensive plated zinc fittings and painted steel

**30**. *This SA 1935 Honour Dagger is embellished by the addition of a fine Damascus steel blade and a leather covering to the scabbard.*

scabbards. Production of the *SA* dagger ceased in 1943, by which time over 3 million had been manufactured.

## THE *SA* 1934 RÖHM HONOUR DAGGER

The purchase of a dagger constituted a considerable financial commitment for many early *SA* men, a sizeable proportion of whom were unemployed and supporting large families. To ease this burden on his longest-serving 'old guard', *Stabschef* Röhm issued an order on 3 February 1934 granting free daggers to all 135,860 *SA* officers and men who had been members of the organisation, or the Hitler Youth, before 31 December 1931. These complimentary pieces were etched on the blade reverse with the dedication *In herzlicher kameradschaft, Ernst Röhm* ('In heartfelt comradeship, Ernst Röhm'), in an imitation of Röhm's handwriting. Otherwise, they were identical in design to the standard *SA* service dagger.

On 21 February 1934, the *Stabschef* directed that his free daggers should be known as *Ehrendolche*, or Honour Daggers, and he permitted other *SA* commanders to present similar weapons, suitably etched or engraved, to their own favoured colleagues locally on special occasions. Röhm himself took to wearing a unique chained dagger featuring an ornate scabbard and both the *SA* and the *SS* insignia on the grip.

After the fall of Röhm in the Night of the Long Knives, it was ordered on 4 July 1934 that the Honour Daggers bearing the *Stabschef's* dedication should be surrendered and replaced by standard *SA* service daggers. However, the Honour Daggers could continue to be worn provided the offending inscription was removed by grinding. This led to several different variations of the Röhm dagger being carried in that some *SA* men had the entire dedication removed professionally, while others crudely filed it off themselves and some obliterated only Röhm's signature, leaving the rest of the etching untouched. A small minority ignored the order altogether and retained their Honour Daggers intact in memory of the popular *Stabschef*, who had successfully nurtured something of a

personality cult amongst certain sections of his followers.

The Röhm *Ehrendolch* was basically nothing more than a 'free gift', acknowledging long *SA* service. However, it soon started a fashion amongst other Nazi organisations of creating similar Honour Daggers to be presented on a more restricted and formal basis as awards for meritorious conduct.

## THE *SA* 1935 HONOUR DAGGER

In 1935, Viktor Lutze created a new *Ehrendolch* for bestowal upon *SA* officers and men at his own discretion, in recognition of exceptional service. The weapon was similar in appearance to the

**31**. SA *1938 Honour Dagger presented by* Stabschef *Lutze to the artist and illustrator Kunz Richter, in recognition of services rendered to the* Sturmabteilung. *Richter designed many* SA *uniform accoutrements and drew up posters publicising the organisation.*

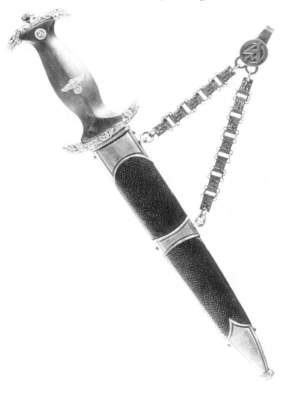

standard *SA* dagger but had high-relief oakleaf ornamentation on the hilt fittings and an engraved border around the scabbard locket and chape. The plain steel blade normally bore a suitable inscription.

### THE *SA* 1938 HONOUR DAGGER

In 1938, Lutze instituted yet another Honour Dagger to be conferred for merit. However, in this case eligibility was restricted to so-called '*SA* high leaders', that is to say officers above the rank of *Standartenführer*. Manufactured by Carl Eickhorn, the new piece was based on a unique dagger previously presented by the *SA* hierarchy to Lutze himself,

and took the form of the 1935 *Ehrendolch* with the addition of an ornate hanging chain bearing raised gold swastikas and the *SA* runic insignia. The blade was of Damascus steel, while the scabbard was covered in brown leather. It was seldom bestowed.

### THE *SA FELDHERRNHALLE* DAGGER

Created in 1935, the *SA-Standarte Feldherrnhalle* acted as Lutze's ceremonial bodyguard and was the elite formation of the entire *SA* organisation. In

**33**. *Hermann Göring wearing his own unique version of the* SA *Honour Dagger, with its distinctive elongated chains, at Nuremberg in September 1937.*

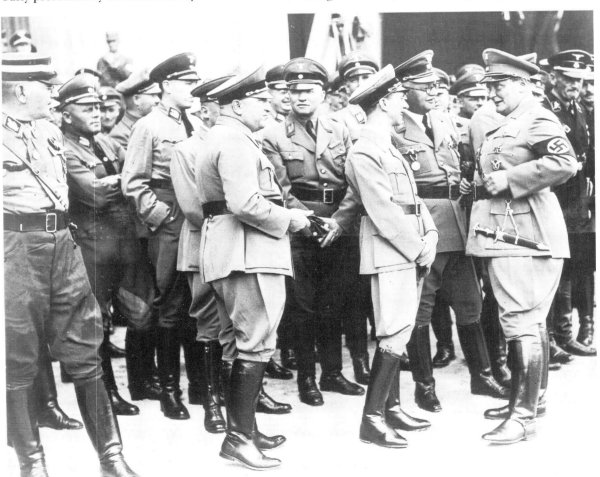

32. *Close-up detail of the elaborate fittings which marked the SA 1938 Honour Dagger as one of the most ornate edged weapons produced during the Third Reich.*

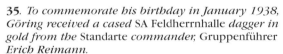

35. *To commemorate his birthday in January 1938, Göring received a cased SA Feldherrnhalle dagger in gold from the* Standarte *commander,* Gruppenführer *Erich Reimann.*

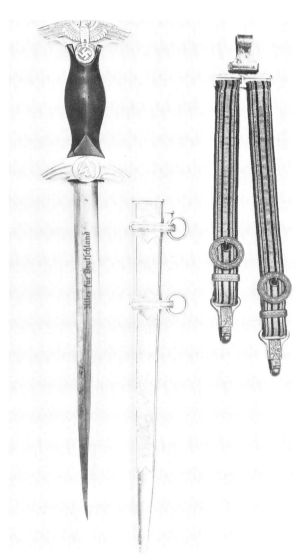

34. *The SA Feldherrnhalle dagger, fewer than ten examples of which are known to survive in various states of disrepair. This piece is complete, and accompanied by its ultra-rare regulation hanging straps.*

1937, a special dagger designed by Paul Casberg was produced by Carl Eickhorn and distributed to *Feldherrnhalle* officers and selected members of Lutze's personal staff. The 45 cm weapon featured a plain brown *SA*-style grip with an eagle pommel

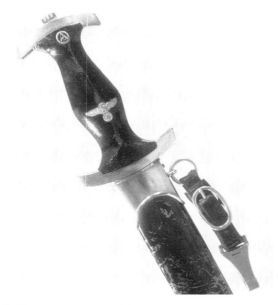

**36**. *The early version of the Naval* SA *dagger, with black grip and scabbard, was worn until 30 August 1934.*

and the *SA* runes on the crossguard. Its long, narrow blade was etched with the traditional motto *Alles für Deutschland*, while the scabbard was in plain aluminium. All metal fittings were silvered. The dagger was worn from a brown velvet-backed fabric hanger faced with gold brocade and narrow brown stripes.

A gilded version of the *Feldherrnhalle* dagger was presented to Lutze on 28 December 1937, his forty-seventh birthday. He wore it with a dress knot bearing the *SA* monogram. Hermann Göring, Honorary Colonel of the *Feldherrnhalle* regiment, received another golden example, but with a white grip, on 12 January 1938.

Reputedly, fewer than 100 examples of the *Feldherrnhalle* dagger were ever produced. However, it is noteworthy that in issue no. 47/1938 of the official *SA* newspaper *Der SA Mann*, the dagger is described as the '*SA Führerdolch Modell 1937*', with no reference whatsoever to the *Feldherrnhalle Standarte*. It is probable that wider distribution of the new dagger, which was akin to

those of *Wehrmacht* officers, was initially contemplated at a time when plans were afoot for sizable elements of *Feldherrnhalle* to be incorporated into the Army and *Luftwaffe*. It is not impossible that the new dagger was intended for officers of a '*Waffen-SA*' force which never ultimately evolved.

## NAVAL *SA* DAGGERS

Each *SA Gruppe* had within its organisational structure at least one naval company, or *Marine-Sturm*, which comprised *SA* men who had had First World War naval experience or who were professional sailors, fishermen or similar. Among other tasks, they recruited and trained their land-based colleagues for service in the *Kriegsmarine* or merchant navy. From 1934, members of the Marine *SA* wore a navy blue uniform. To complement this, they were issued with a variant of the *SA* service dagger which had a black grip and scabbard. After a short period, the black dagger was replaced by a brown one with distinctive gilded metal fittings. From 1936, naval units reverted to wearing the standard *SA* service dagger in brown and silver.

## *SS*

Originally part of the *SA* organisation, the *Schutzstaffeln* (protection squads) began life as Hitler's personal bodyguard and developed rapidly after the Nazi consolidation of power in 1933. New *SS-Verfügungstruppe* were formed as barracked quasi-military forces to bolster the regime in the event of political turmoil or counter-revolution, while the *SS-Totenkopfverbände* were recruited to guard the growing number of concentration camp inmates. The bulk of the general *SS*, or *Allgemeine-SS*, remained part-time and unpaid, and membership included the entire spectrum of the German civil population, from farm labourers to landed aristocracy and from small shopkeepers to the directors of national companies. Its academics engaged upon research projects and intelligence-gathering activities which had a great influence on both the domestic and foreign policies of the Third Reich. The

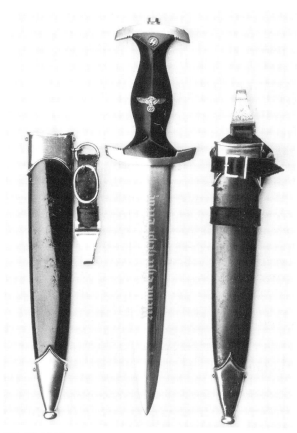

**38**. *The* SS *blade motto, 'My Honour is Loyalty'.*

Much more has been written about the *SS* than any other Nazi organisation, so further coverage here is unnecessary.

## THE *SS* 1933 DAGGER

The *SS* service dagger was introduced along with its *SA* counterpart on 15 December 1933, and it was very similar in design. Black and silver in colour, it bore the *SS* motto *Meine Ehre heisst Treue* ('My Honour is Loyalty') on the blade and *SS* runes on the upper part of the grip. Worn by all ranks of the *Allgemeine-SS* with service and walking-out dress, the *SS* dagger was given a special status and was presented to its owner only at the annual 9 November ceremony when he graduated from *SS-Anwärter*, or candidate, to a fully confirmed *SS-Mann*. It was not issued at any other time, or *en masse* like the daggers of the plebeian *SA*. Each *SS* member paid the full cost of his dagger, usually in small instalments, prior to its presentation.

On 17 February 1934, *SS-Gruppenführer* Kurt Wittje, Chief of the *SS-Amt*, forbade the private purchase or 'trading in' of *SS* daggers on the open market. Henceforth, daggers could be ordered only through *SS* headquarters, for issue via the three main *SS* uniform distribution centres at Munich, Dresden and Berlin. These centres were responsible for carrying out quality inspections of *SS* daggers, after which their respective control marks of 'I', 'II' or 'III' were stamped into the crossguard reverses. It was subsequently made a disciplinary

**37**. SS *1933 Dagger, showing the single leather strap hanger (left) and vertical hanger (right). Note the slanted position of the runes badge, typical of 1934–5 pieces. This example was produced by Gottlieb Hammesfahr and the crossguard reverse is stamped with the 'III' inspection mark of the Berlin SS uniform distribution centre. Most, if not all, Hammesfahr SS daggers appear to have been processed through Berlin.*

Government, industry, commerce, education and particularly the police all came to be dominated by the *SS* under its *Reichsführer*, Heinrich Himmler. By the autumn of 1944, through a well-planned and executed strategy of infiltration, the *SS* had achieved almost total political, military and economic control of Germany, and Himmler was widely regarded as Hitler's natural heir-apparent.

offence for an *SS* man to dispose of or lose his dagger, on the grounds that it was a symbol of his office. In these ways, it was assured that no unauthorised person could buy or otherwise acquire an *SS* dagger. As of 25 January 1935, members dismissed from the *SS* had to surrender their daggers. In cases of voluntary resignation or normal retirement, however, daggers could be retained and the man in question was given a certificate stating that he was entitled to possess the dagger.

The *SS* dagger was suspended at an angle from a single leather strap until November 1934, when Himmler introduced a vertical hanger for use with service dress during crowd control. However, while the vertical hanger was more stable it was too reminiscent of the humble bayonet frog, and in 1936 the single strap was reintroduced for both the walking-out and service uniforms. Thereafter, the vertical hanger was restricted to use on route marches and military exercises. No portepée knot was worn with the *SS* 1933 dagger.

From 1936, the production and supply of *SS* daggers was controlled by the *RZM*, with a consequent phasing out of the old *SS* inspection stamps. As with the *SA-Dolch*, post-1936 *SS* daggers were generally lighter in weight, with painted rather than anodised scabbards and less expensive plated fittings.

In September 1940, due to national economies, the *SS* 1933 dagger was withdrawn from production for the duration of the war.

### THE *SS* 1934 RÖHM HONOUR DAGGER

An *SS* version of the *SA* Röhm Honour Dagger was distributed by the *Stabschef* to 9,900 *SS* men who met the same qualifying criteria as those laid down for members of the *SA*. The dagger was identical in design to the standard *SS* service dagger, but had the blade reverse etched with the dedication *In herzlicher kameradschaft, Ernst Röhm*. On 12 April 1934, *SS-Gruppenführer* Wittje ordered that each *SS* Röhm Honour Dagger should be marked on the crossguard reverse with the *SS* membership number of the recipient, who was also to be given a certificate confirming his entitlement to the dagger. Immediately after the Night of the Long Knives, *SS* holders of the

**39**. SS *1934 Röhm Honour Dagger by Eickhorn, with intact dedication. The recipient complied with the instruction to stamp his* SS *membership number '21779' on the crossguard reverse, but apparently ignored the order to remove the offending Röhm inscription.*

Röhm Honour Dagger were, like their *SA* comrades, ordered to remove the offending inscription. The vast majority had no hesitation in doing so.

### THE *SS* 1934 HIMMLER HONOUR DAGGER

The Night of the Long Knives marked a turning point in the history of the *SS*. Following the murder of Ernst Röhm, the organisation was removed from

**40**. SS *1934 Himmler Honour Dagger, with blade dedication in imitation of the 34-year-old* Reichsführer's *distinctively angular script.*

**41**. *This SS 1936 chained dagger is decorated in the regulation manner with the portepée knot authorised for officers of the* Waffen-SS *and Security Police in 1943.*

the control of the *SA* and firmly established as an independent formation of the *NSDAP*. Himmler was simultaneously promoted and made answerable only to Hitler. On 3 July 1934, the *Reichsführer-SS* authorised 200 new Honour Daggers for those of his men who had participated most fully in the bloody purge of the *SA*. Each piece was identical to the standard *SS-Dolch* but had the blade reverse etched with the dedication *In herzlicher kameradschaft, H. Himmler*, in an imitation of Himmler's handwriting. Each recipient's *SS* membership number was stamped on the back of the crossguard. The sole manufacturer of this rare piece was Carl Eickhorn.

## The *SS* 1936 Chained Dagger

A more ornate *SS* dagger, to be worn only by officers and by those old guard NCOs and other ranks who had joined the organisation prior to 30 January 1933, was introduced by Himmler on 21 June 1936. Generally known as the chained dagger, it was similar to the basic *SS* service dagger but was suspended by means of linked octagonal plates finely embossed with death's heads and *SS* runes,

**43**. SS *personnel engaged in crowd control during a parade in Berlin, 1939. Of particular interest is the fact that the officer in the centre has, like his men, suspended his chained dagger in a vertical position behind his back.*

**42**. *An 'Old Guard'* SS-Untersturmführer *from the* Landshut Standarte *poses with his chained dagger in 1937. His decorations indicate service with Austro-Hungarian forces during the First World War.*

**44**. *The reverse of* SS *1936 dagger chains. The lower right link shows clearly the entwined Sig-runes inspection stamp indicating that the design had been commissioned and approved by the* SS, *according to its own strict artistic standards. While* SS *1936 daggers were themselves unmarked, it is known that all their chain fittings were produced under subcontract by the Munich jewellery firm of K. Gahr, which was patronised heavily by the* SS *and NSDAP.*

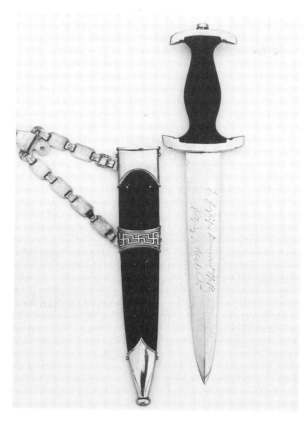

46. *The* SS *1936 Honour Dagger was manufactured exclusively by Carl Eickhorn. Only two surviving examples of this variant, marked by elaborate raised oakleaf decoration to the scabbard mounts, are known.*

45. *From 21 February 1934, senior* SS *commanders were authorised to commission production of their own Honour Daggers for limited presentation to deserving personnel on a local basis. This* SS *1936 chained dagger was awarded by* Obergruppenführer *Josias* Erbprinz zu Waldeck-Pyrmont, *commander of* SS-Oberabschnitt 'Fulda-Werra', *and bears the dedication* In herzlicher kameradschaft, Erbprinz zu Waldeck *('In heartfelt comradeship, the Prince of Waldeck') in imitation of his handwriting.*

and featured a central scabbard mount decorated with swastikas. The reverse of the chain was stamped with the *SS* proofmark of two entwined Sig-runes, and the suspension clip connector took the form of an ancient Germanic motif known as Wotan's knot. During the 1936–7 period, the chain and fittings, which were designed by Karl Diebitsch, were made from nickel silver. Later examples were in nickel-plated steel, with slightly

smaller, less oval-shaped skulls. Chained daggers bore no makers' marks and it is likely that only one firm, probably Carl Eickhorn, was contracted to produce them.

Each chained dagger, which cost 12.15 Reichsmarks, had to be purchased through official channels, and could initially be worn only with the black uniform. On 15 February 1943, however, *Waffen-SS* officers were permitted to sport the weapon with their field-grey walking-out dress and were allowed to attach portepée knots in the Army style, as a concession to their military status. They were also granted authority to buy their daggers direct from *RZM*-approved shops. The same rights were afforded to officers of the Security Police and the *Sicherheitsdienst* (*SD*) from 4 June 1943. Production of the chained dagger had to be discontinued at the end of that year because of material shortages, and its wear was subsequently forbidden for the duration of the war.

## THE *SS* 1936 HONOUR DAGGER

A very ornate *SS* Honour Dagger, with oakleaf-decorated crossguards, leather-covered scabbard and Damascus steel blade, was created by Himmler in 1936 for presentation to high-ranking officers in recognition of special achievement. It was normally suspended from the standard single-strap hanger. When one was bestowed upon the *NSDAP* Treasurer, *SS-Oberst-Gruppenführer* Franz Xaver Schwarz, he responded by secretly commissioning the Eickhorn firm to produce an even more elaborate example, with a chain hanger and fittings in solid silver, which he then gave to Himmler as a birthday present.

## THE *WAFFEN-SS* DAGGER

Hitler intended that the *SS* would eventually police his New Order in Europe. To give them the moral

**47**. *Drawing for a prototype* Waffen-SS *dagger, produced by the Peter Krebs firm during May–June 1940.*

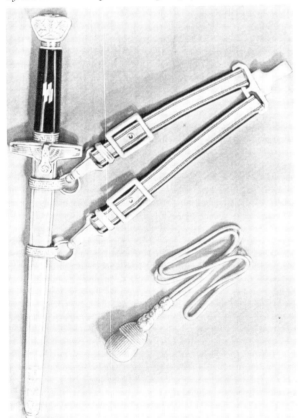

authority to do so, it was essential that they should first 'win their spurs' on the battlefield, in their own closed units. To that end, the *SS-Verfügungstruppe* and *SS-Totenkopfverbände* participated in the Polish campaign and, during 1939–40, amalgamated under the new title of *Waffen-SS*. The *Waffen-SS* subsequently took part in most of the major battles of the war, and by 1945 had expanded to include over 30 divisions comprising almost 1 million troops.

On 24 May 1940, *SS-Obergruppenführer* Fritz Weitzel suggested to Himmler that an Army-style dagger should be created for wear exclusively by officers of the *Waffen-SS*, who had performed so well in combat but were prevented by regulations from wearing existing *SS* daggers with their field-grey uniform. Three trial designs for a *Waffen-SS* dagger, produced by the Solingen firms of Alexander Coppel and Peter Krebs and by the *SS* Damascus School at Dachau, were duly submitted for the *Reichsführer*'s consideration but were rejected out of hand. Weitzel was killed during an air raid on Düsseldorf on 19 June 1940, and his project was subsequently shelved.

# NATIONAL SOCIALIST MOTOR CORPS

In 1922, the Nazis acquired a small fleet of cars, lorries and motorcycles which they used very effectively for the purpose of transporting *SA* men and propaganda material to political meetings. Ordinary party members were subsequently encouraged to enhance this facility by lending their own private vehicles to the *NSDAP* as and when the occasion demanded. However, it became clear that the *ad hoc* nature of such an arrangement reduced its effectiveness, so on 1 April 1930 the *Nationalsozialistisches Automobilkorps*, or *NSAK*, was formed, mobilising all Nazi car owners and driving enthusiasts into formalised motoring units. Adolf Hühnlein, an ex-Army major and early associate of Hitler with no mechanical knowledge whatsoever, took over the Corps at the end of 1930 and expanded it into uniformed *Motorstandarten* organised along *SA* and military lines.

assume responsibility for providing the *Napola*s with their clothing and equipment, and also began to sponsor scholarships for ethnic German students from abroad. On 9 March 1936, the *SS* commitment to the schools was rewarded with the appointment of *SS-Obergruppenführer* August Heissmeyer as Inspector-General of the *NPEA*. By 1940, in a prime example of its assumption of power and influence through a steady policy of infiltration, the *SS* had completely taken over the *Napola*s, with full authority in matters relating to curriculum and staff appointments. *NPEA* commandants and teachers were subject to the *SS* discipline code, which meant that they owed unswerving loyalty to Himmler.

The rhythm of *Napola* life was thereafter based on that of Himmler's Black Order. Conventional religion was abolished from the curriculum and replaced by the study of pagan Germanic rites. The celebration of *Julfest*, the *SS* Christmas, brought the pupils together to worship the Child of the Sun, arisen from his ashes at the Winter Solstice. New school songs commemorated the struggle between day and night, and praised the eternal return of life. The night of 21 June became the Night of the Sun, when the boys mounted a 'Joyous Guard' awaiting the sun's triumphal reappearance. Lectures were given on racial superiority and *SS* ideology, and emphasis was placed on duty, courage and personal obligation.

The first all-girls' *NPEA* school was opened in 1941 at Achern in Baden, and some of the previously all-male establishments subsequently admitted female pupils and staff. No fewer than twenty-seven new *Napola*s were founded between 1941 and 1942, and with the enormous expansion of the *NPEA* programme during the war *SS* influence became paramount. For example, three schools, known as *NPEA Reichsschulen*, were set up in the occupied western territories specifically to take in non-German Nordic pupils, the future leaders of a pan-European *SS*.

In December 1944, by virtue of his success with the *NPEA*s, and his position as Commander-in-Chief of the Army Reserve, Himmler was appointed by Hitler to be supervisor of all schools from which

future *Wehrmacht* and *Waffen-SS* officers could be recruited. In theory, that put him in charge of almost every educational establishment in the Third Reich and conquered countries!

The *SS* influence on the *NPEA* was most apparent in its dress. They systematically introduced new uniforms and a scheme of ranks, titles and insignia which were entirely *SS*-based. Amongst these accoutrements were daggers for both staff and students.

## THE *NPEA* 1935 STUDENT'S DAGGER

Junior pupils in their first four years at *NPEA* schools always wore the basic Hitler Youth knife

**54**. *An* NPEA *1935 student's dagger, with aluminium fittings. The upper and lower hilt mounts are engraved with the property designations 'NPEA' and 'Oranienstein', respectively.*

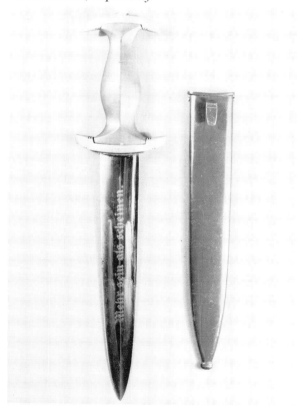

55. *Most* NPEA *daggers bore the etched mark of their distributor, 'Karl Burgsmüller, Berlin', rather than a maker's logo.*

56. *An* NPEA *student's dagger, stamped on the lower crossguard with the typical property designation 'N 119', denoting the 119th dagger on the inventory of the* Napola *School at Naumburg/Saale. The engraving on the upper grip mount, which reads 'N.P.E.A. Naumburg – Weihn. 35', suggests that this particular piece was gifted by staff as a souvenir to a student graduating at Christmas 1935, less than two months after the* NPEA *dagger was introduced. The regulation suspension frog is also shown clearly.*

with *Napola* uniform. Towards the end of 1935, a new Holbein-style dagger was authorised for senior students only, those between the ages of fourteen and eighteen. Much plainer than the weapons of the *SA*, *SS* and *NSKK*, it had a bare brown wooden grip, without insignia, and was held in an unadorned olive-green steel scabbard of the bayonet variety, suspended vertically from a simple brown leather frog. The blade bore the *NPEA* motto *Mehr sein als scheinen*. Hilt fittings were initially of nickel silver, but after 1936 they were produced in aluminium. While *NPEA* daggers were manufactured by Carl Eickhorn and Max Weyersberg, they seldom bore these makers' logos. More commonly, they featured the trade mark of Karl Burgsmüller of Berlin, the wholesale distributor from whom the *Napola*s obtained their supplies.

*NPEA* daggers normally remained school prop-

erty and were issued to pupils on a temporary basis, to be returned when their studies were completed. Consequently, each piece was stamped on the crossguard with an abbreviation denoting the originating *Napola* establishment, and an accountability number. For example, 'WT23' signified the twenty-third dagger on the inventory of the *NPEA* school at Wien-Theresianum. The following table lists all the *NPEA*s whose details may be found inscribed on daggers, with the abbreviations known to have been used:

| NPEA | ABBREVIATION |
|------|--------------|
| Anhalt | A |
| Backnang | Ba |
| Bensberg | B |
| Berlin-Spandau | Sp |
| Ilfeld | J |
| Klotzsche | Kl |
| Köslin | K |
| Loben | L |
| Naumburg/Saale | N |
| Neuzelle | Ne |
| Oranienstein | O |
| Plön | Pl |
| Potsdam | P |
| Potsdam Grosses Waisenhaus | PGW |
| Reisen | Re |
| Rottweil | R |
| Rufach | Ru |
| Schulpforte | Sch |
| Stuhm | St |
| Traiskirchen | T |
| Wahlstatt | W |
| Wien-Breitensee | WB |
| Wien-Theresianum | WT |

A number of other *Napola*s were founded after 1942, but the production of *NPEA* daggers ceased around that time so these later institutions did not issue edged weapons to their pupils.

On occasion, graduating students were presented with daggers as permanent mementoes of their time with the *NPEA*. Such pieces were usually stamped with a suitable inscription denoting the graduation date.

### THE *NPEA* 1935 STAFF DAGGER

A dagger was also created in 1935 for teachers and full-time members of the directing staff at *NPEA* establishments. It was identical to the student's

*57. An NPEA 1935 staff dagger, by Eickhorn. Note the long Gothic-style 'S' letters on the blade motto, characteristic of Eickhorn-made pieces.*

dagger, but featured an eagle and swastika on the grip.

### THE *NPEA* 1936 STAFF CHAINED DAGGER

During the autumn of 1936, the *NPEA* staff dagger was upgraded by the addition of a plain double chain hanger comprising nickel silver or aluminium

**58**. *An* NPEA *1936 staff chained dagger.*

rings, identical to the type used with the *Luftwaffe* 1935 dagger. A scabbard similar to that of the *NSKK* 1936 chained dagger, but painted olive-green, was employed to accommodate this accessory. From the outset it was decided that the small quantity required necessitated extreme economy in production, hence the use of existing and plentiful *NSKK* and *Luftwaffe* dagger component parts.

Even as late as 1942, when *Napola* dagger manufacture was suspended, there were only 500 permanent teaching staff in the entire *NPEA* system. This compared with 10,000 *SS* officers listed that year. Given their very limited numbers and the fact that *Napola* teachers were obliged to share daggers 'on loan' from their various establishments, it is clear why the *NPEA* chained dagger was one of the rarest of all Third Reich edged weapons.

Some *NPEA* daggers survive which have been married up with standard-issue brown or black *SA*, *SS* or *NSKK* scabbards. While most *NPEA* staff also held membership of other *NSDAP* organisations, particularly the *SS*, there is no evidence to suggest that the carrying of *NPEA* daggers in non-*NPEA* scabbards was a contemporary custom.

## HITLER YOUTH

While the majority of ordinary German youngsters never had any associations with the *NPEA*, most either belonged to or had friends in the Hitler Youth (*Hitlerjugend* or *HJ*) and its female equivalent, the League of German Girls (*Bund Deutscher Mädel,* or *BDM*). After 1933 the *HJ* was a main source of recruitment for the *Allgemeine-SS*, and as the power and prestige of the *SA* declined so those of the *SS* and *Hitlerjugend* increased. In 1936 it was decreed that the whole of German youth was to be educated, outside the parental home and school, in the *HJ*, physically, intellectually and morally, for service to the nation and community. The Hitler Youth initially found it hard to meet the great demands placed upon it, and for that reason obligatory membership was delayed for several years. Even so, voluntary enlistment resulted in the membership reaching 8 million (i.e. 66 per cent of those eligible to join) at the end of 1938. Compulsory *HJ* service for all male 17-year-olds was introduced on 25 March 1939, and in September 1941 membership finally became obligatory for both sexes from the age of ten onwards. Many of the activities, trappings and insignia of the *HJ* were derived from those of the *SS*, and co-operation between the two organisations became ever closer until by the end of the war they had merged their interests completely, and the ultimate aim of every Hitler Youth was acceptance into the *SS*.

Baldur von Schirach was appointed *Reichsjugendführer* in 1931 and led the *HJ* during its years of greatest expansion. He was rewarded for his efforts by being nominated *Gauleiter* of Vienna in August 1940, at the age of only 33, and was succeeded as Reich Youth Leader by Artur Axmann. From 1936, the *HJ* ran weekend courses in field exercises and rifle shooting, using *Wehrmacht* and

SS instructors. In 1939, military training camps were set up in which boys between the ages of sixteen and a half and eighteen were put through a three-week course culminating in the award of a War Training Certificate, or *K-Schein*. The *Waffen-SS* possessed no powers of direct conscription amongst German nationals, but if a young man could be persuaded to volunteer for the *Waffen-SS* before reaching his twentieth year, the normal age for conscript service, his preference for that branch of the fighting forces was normally respected. The *SS* therefore strove to persuade *HJ* boys to volunteer for service in one of its combat divisions after they had obtained their *K-Schein*.

In February 1943, following the loss of the 6th Army at Stalingrad, manpower shortages became so acute that Hitler authorised a programme to encourage voluntary enlistment of 17-year-olds, boys who would not have been subject to conscription until 1946. The *SS* saw this as a golden opportunity to build up its own forces. Negotiations between Himmler and Axmann began at once, as a result of which it was decided to raise an entirely new *Waffen-SS* division from Hitler Youths who had attained their War Training Certificates. By midsummer the required number of 10,000 volunteers had been mustered and in October the division was officially named the 12th *SS*-Panzer Division '*Hitlerjugend*'. It subsequently fought well in Normandy and Hungary, and by the end of the war had been reduced to a single tank and 455 men.

Practical use was also made of *HJ* members on the home front. By the middle of 1943, there were some 100,000 young Germans in the Auxiliary Flak Organisation, run by the *Luftwaffe*. They served primarily as anti-aircraft gunners and searchlight operators. *HJ-Feuerwehrscharen*, or Fire Defence Squads, accounted for some 700,000 boys engaged in firefighting associated with air raids, and during the winter of 1944–5 Hitler Youths in the *SS*-led *Wehrwolf* guerrilla organisation harassed Allied troops advancing into Germany. From its 'Boy Scout'-type origins in the 1930s, the Hitler Youth had developed into a military and political force to be reckoned with.

### THE *HJ* 1933 KNIFE

The Hitler Youth was the first Nazi organisation to carry an official edged weapon. During the 1920s, a large number of independent scouting-type groups such as the Pathfinders and the Good Comrades existed under the auspices of the Reich Committee for German Youth Associations, and members of these bodies commonly utilised small travelling knives, or *Fahrtenmesser*, on their hiking and camping trips. In 1928, Hitler directed that his own *HJ* boys should likewise furnish themselves with similar equipment, although no specific blade pattern was prescribed. For the next five years, therefore, the Solingen makers happily supplied Hitler Youths with a wide variety of knives, generally taking the form of surplus First World War short-bladed trench daggers with the addition of swastika insignia. Some pieces were more elaborate than others, incorporating staghorn grips and ornately etched hilts or blades, depending upon the amount of money each individual boy or his parents could afford to spend in these times of widespread economic depression.

*59. A Hitler Youth 1933 knife, with the blade etching which was discontinued on 28 August 1938.*

In February 1933, von Schirach decided that a degree of uniformity was required and he duly published regulations standardising the pattern of *Fahrtenmesser* to be worn by all *HJ* and *Deutsche Jugend* (*DJ*) officers and boys upon completion of their probationary training. The introductory order stated:

> A sheath knife is being created for our youngsters which will be useful to them on hiking excursions and so on. However, it will also serve as a means of impressing upon them that an edged weapon has great traditional significance and that the bearer of such a weapon has a heavy responsibility which accompanies his right to carry it.

The small but functional Hitler Youth knife, which retailed at 4 Reichsmarks, had a nickel-plated steel hilt with a chequered black bakelite grip into which was set the *HJ* diamond badge in red, white and black enamel.

Its black stove-enamelled steel scabbard was suspended from a simple belt loop hanger with press-stud fastener. Initially, the 14 cm blade was etched with the Hitler Youth motto *Blut und Ehre!* ('Blood and Honour!') in an imitation of von Schirach's handwriting. However, this feature was discontinued in August 1938 following distribution of the leader's dagger.

Production ceased in 1942, by which time over 15 million examples of the *HJ* knife had been manufactured, making it by far the most mass-produced dress weapon of the Third Reich. A small number were modified for members of the National Socialist Students' League and foreign pro-Nazi youth movements by the incorporation of appropriate insignia in the grip, but these appear to have been privately ordered and were never officially sanctioned.

## THE *HJ* LEADER'S 1937 DAGGER

It soon became clear that the *HJ-Fahrtenmesser*, while a very practical piece of camping equipment for boys, was not sufficiently impressive to grace the increasingly flamboyant uniforms of adult

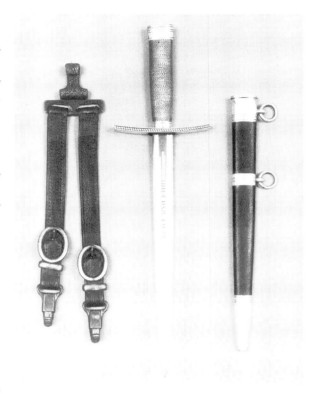

**60**. *An* HJ *leader's 1937 dagger, by E. & F. Hörster.*

**61**. *Even after being appointed* Gauleiter *of Vienna in 1940, Baldur von Schirach (centre) continued to wear his* HJ *leader's dagger with* NSDAP *uniform.*

Hitler Youth officers, especially during ceremonial occasions. Consequently, in 1937 von Schirach introduced a slim 35 cm stiletto dagger to be worn by selected full-time *HJ* leaders ranking from *Stammführer* upwards. The new *Führerdolch* had the *HJ* emblem on the top of the pommel, a wooden grip tightly bound with silver wire, and a simple grooved design on the plain upward-sloping crossguard. Its blade was etched with the *Blut und Ehre!* motto, in upper-case Latin lettering. The scabbard featured silver-plated fittings with the *HJ* flighted eagle on the locket, and was bound in black leather. The weapon was suspended from two black leather straps, and was carried without a knot.

Wearing of the *Führerdolch* was restricted to deserving individuals specifically authorised by the *Reichsjugendführer*, and it was presented in a case together with an official certificate. It could not be bought at retail outlets. Only two firms, Carl Eickhorn and E. & F. Hörster, were contracted by the *RZM* to produce it.

In 1938, during the highly publicised visit of a Hitler Youth delegation to Japan, some thirty *HJ* officers wore a special version of the dagger with a plain silver scabbard and white leather hanging straps. A variant with gilded fittings was also manufactured, possibly for bestowal upon *HJ* generals of the rank of *Obergebietsführer* and above, or upon officers of the *Marine-HJ*, to match the gold insignia on their uniforms.

Production of the elegant *Führerdolch* ceased in October 1942.

# GERMAN AIR SPORTS ASSOCIATION AND NATIONAL SOCIALIST FLYING CORPS

The National Socialist Flying Corps, or *Nationalsozialistisches Fliegerkorps* (*NSFK*), was founded in January 1932 to attract air-minded Germans into the *NSDAP* and to stress the role of aviation in modern society. In March 1933, it was absorbed into a new German Air Sports Association, the *Deutscher Luftsport Verband* (*DLV*), which Hitler had established with a view to training recruits for

his planned air force. The *DLV* grew dramatically, swallowing up all existing flying clubs throughout the country and securing their airfields, gliders and other equipment. *DLV* personnel were kitted out in a uniform unashamedly based on that of the Royal Air Force, and by the end of 1934 had come to be regarded as military auxiliaries. When the *Luftwaffe* was officially inaugurated as an independent branch of the *Wehrmacht* on 26 February 1935, the purpose of the *DLV* had been accomplished and it was finally dissolved in July of the following year.

On 17 April 1937, the *NSFK* was resurrected and placed under the command of *Luftwaffe* General Friedrich Christiansen, who enjoyed a close friendship with Göring. Never a very large organisation, it assumed the role of a 'state flying club' and provided training in the piloting and maintenance of powered aircraft, gliders and balloons. It also gave specialist instruction in wireless operation and parachuting, and co-operated with the Hitler Youth to encourage boys of the *Flieger-HJ*. By 1939, the *NSFK* was producing a constant flow of skilled recruits for the *Luftwaffe* and thereby functioned as a reserve pool for Göring's forces.

The history and functions of the *DLV* and *NSFK* were therefore closely aligned, and this was reflected in the similarity of their edged weapons.

## THE *DLV* 1934 DAGGER

In March 1934, *DLV-Präsident* Bruno Loerzer heralded the creation of a dagger for his officers in the following terms:

> The new *Fliegerdolch* is distinguished by its frugal ornamentation, the classic design impressing by virtue of its simplicity. It embodies the new spirit of our era of linearity and plainness.

Medieval in appearance, the weapon was almost identical to the later *Luftwaffe* 1935 dagger, which was based upon it. However, the *DLV* dagger was much larger and heavier, measuring 55 cm in length, by far the longest of all standard Nazi daggers. Manufactured by Carl Eickhorn and Paul

**63**. *A* DLV-Flieger-Vizekommodore *(second from right) wearing the lengthy* Deutscher Luftsport Verband *dagger on Göring's forty-second birthday, 12 January 1935.*

**62**. *A* DLV *1934 dagger by Paul Weyersberg. Of particular note are the leather grip, which is devoid of any wire wrapping, and the staples securing the scabbard fittings in place. The hanging clip bears the pattern number 'DRGM 926719'.*

Weyersberg, its wooden grip was wrapped in dark blue leather, while all metal fittings were of nickel silver. The leather scabbard was suspended from a plain double chain hanger and had its metal portions secured by staples rather than the usual screws. Examples of the *Fliegerdolch* produced by Eickhorn had a distinctive square quillon block, while those made by Weyersberg had the normal round type. The dagger was worn with a 42 cm silver bullion portepée knot. Production of this short-lived model ceased in March 1935.

## The *DLV* 1934 Knife

At the same time as the *DLV* dagger was introduced, a simpler and shorter knife was authorised for all non-officer ranks of the *DLV*. The so-called *Fliegermesser*, or 'airman's knife', measured 34 cm in length and featured an unadorned wooden grip and scabbard covered in dark blue leather. All metal fittings were again in nickel silver. The crossguard took the form of downward-sloping wings bearing a central black enamel swastika, and the knife was suspended from a single blue or brown leather strap hanger permanently attached to the carrying ring of the upper scabbard fitting. The throat of the scabbard was usually stamped with the *DLV* emblem, comprising a winged swastika and propeller, indicating that the knife had passed inspection at the *DLV* establishment responsible for issuing it. No portepée knot was worn with the weapon. From 1936, the *Fliegermesser* was manufactured with cheaper aluminium fittings and with blue enamel paint replacing leather on the grip and scabbard. Upon the final dissolution of the *DLV* in July that year, the knives on issue were recalled and returned to central storage.

1940, when they received their own dagger, which was identical to that used by other civil servants but with a black-toned scabbard. Very few examples were ever produced.

### THE GOVERNMENT GENERAL 1940 DAGGER

In 1940, senior ministerial officials assigned to Hans Frank's semi-autonomous *Generalgouvernement* in Poland were provided with a gold-coloured variant of the government administration dagger, hung from gilt brocade straps and decorated with a golden-yellow wire portepée knot. Intended to set its wearers apart from Reich ministry staff, it was distributed in very limited numbers until mid-1942.

### THE EASTERN MINISTRY 1942 DAGGER

On 25 March 1942, an entirely new pattern of dagger in the imperial Russian style was approved for peacetime wear by higher-grade officials of Alfred Rosenberg's *Ostministerium*, to complement their golden-brown uniforms. Prototype design sketches indicate that it was to have been 41 cm in length, gold in colour, with a concave dished pommel and white plastic grip. The proposed crossguard bore a large Nazi eagle with folded wings on the obverse, and an enamel shield in the national colours of the conquered territory concerned (Estonia, Latvia, Lithuania, Byelorussia or the Ukraine) on the reverse. The weapon was to be suspended from gilt braid hangers with brown velvet backs, and finished off by a 42 cm gold bullion portepée knot.

Since dagger manufacture generally drew to a close from July 1942, it is not surprising that the Eastern Ministry dagger was never ultimately produced. Consequently, German civil servants in Russia who were already in possession of the standard government administration dagger or the

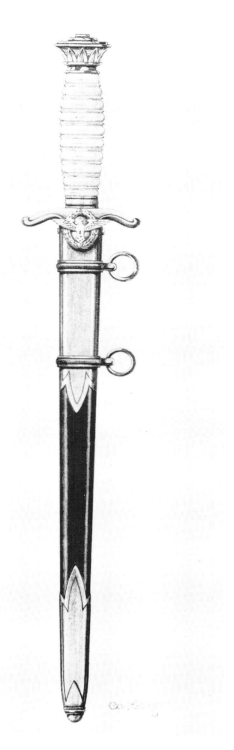

**71**. *A prototype design for a Police officer's dagger, drawn by Paul Casberg for the Eickhorn firm early in 1936 but never approved by the Government. The form of eagle on the crossguard was used by the Police between January 1934 and June 1936.*

Government General dagger continued to wear these instead.

# POLICE

After the failure of the Munich *putsch* in November 1923, when a Nazi attempt to seize power in Bavaria was defeated by the Police rather than the Army, Hitler realised that unrestricted control of the Police would be an essential element in the successful foundation of a long-term Nazi state. Consequently, the period immediately following the *NSDAP* assumption of power on 30 January 1933 witnessed a concerted effort by the *Führer* to have his most trusted lieutenants nominated to senior positions in the various semi-independent provincial police forces which then existed. In January 1934, steps were taken to amalgamate these local forces to form the first German National Police Force, officially termed the *Deutsche Polizei*, and to incorporate the eagle and swastika into the design of Police uniforms. On 17 June 1936, *Reichsführer-SS* Heinrich Himmler was appointed Chief of the German Police and began to formulate the greatest of his many projects, namely the complete merger of the *SS* and Police into a single, all-embracing national State Protection Corps, or *Staatsschutzkorps*. By this means, the conventional and to a degree politically unreliable Police could be done away with altogether. The following years therefore witnessed the nurturing of an ever-closer relationship between the *SS* and the Police, with regular transfers of staff across the two organisations. However, the outbreak of war in 1939 effectively dealt a mortal blow to the steady progression towards a *Staatsschutzkorps* as the majority of its finest young potential recruits were swallowed up by the *Wehrmacht*, with many subsequently being killed or maimed in battle.

Under Himmler, the complex Police network was split into two distinct branches: the Security Police, whose members generally wore civilian clothing; and the Order Police. The latter comprised all uniformed civil Police personnel and included a variety of separate but closely linked formations, each with its own purpose and often its own series of uniforms and accoutrements. These are briefly described below.

The *Schutzpolizei*, or Protection Police, were the regular municipal 'beat bobbies' of the Third Reich and numbered around 200,000 men in 1943. This formation was itself divided into the *Schutzpolizei des Reiches*, whose jurisdiction extended throughout Germany, and the *Schutzpolizei des Gemeinden*, who operated only within their own towns. In addition, companies of *Schutzpolizei* were organised into *Kasernierte-polizei*, or Barracked Police, equipped with armoured cars, machine guns and grenades, and acted as a mobile reserve in emergency situations. The *Gendarmerie*, or Rural Police, covered landward districts and small communities of less than 2,000 inhabitants. They were particularly adept at combating poaching, detecting the black-market slaughtering of animals and the like. Mountain *Gendarmerie*, skilled in climbing and skiing, were employed in alpine areas, and a separate section of the *Gendarmerie*, known as the *Landwacht*, or Rural Guard, was set up in 1942 to supervise prisoners of war engaged on agricultural work. The *Motorisierte Gendarmerie* policed rural roads and the *autobahn* network, while the *Verkehrspolizei*, or Municipal Traffic Police, patrolled town and city streets, enforcing traffic law. Record-keeping, the registration of foreign nationals, the issuing of firearms licences and travel permits and the enforcement of regulations affecting factories and businesses were the prerogatives of the *Verwaltungspolizei*, or Administration Police. Finally, the *Wasserschutzpolizei*, or Waterways Protection Police, were responsible for policing navigable inland rivers and canals, regulating waterborne traffic, preventing smuggling, enforcing safety and security measures and inspecting waterways shipping.

In addition to these groupings, there were a number of other uniformed bodies which performed policing functions but which did not, initially at least, come under the auspices of Himmler's National Police Force. They included the *Bahnschutz*, which maintained order on the German national railway system; the *Postschutz*, which protected post offices; and the *Werkschutz*, which guarded important industrial complexes and factories in general. Such units were operated and paid for by the public corporations or private enterprises concerned.

The majority of Police formations carried bayonets and swords, rather than daggers, as their dress sidearms. These are covered in the relevant chapters.

### THE WATERWAYS PROTECTION POLICE 1938 DAGGER

The *Wasserschutzpolizei*, whose members wore naval uniform with Police insignia and worked closely with the marine and port authorities, was the only branch of the German National Police Force to be issued with a dress dagger. Authorised for officer personnel on 20 April 1938, it was identical to the new Navy dagger introduced at the

*72. The Waterways Protection Police 1938 dagger, clearly based on its naval counterpart, was carried for only a year by a small number of personnel and was one of the rarest of all Third Reich edged weapons. This example was produced by Alcoso.*

same time but retained the flaming ball pommel and featured a dark blue leather-bound grip bearing a gilt metal Police eagle badge. The weapon was suspended from a double strap black velvet hanger with dark blue-edged gilt wire facings, and was worn with a 42 cm gold and blue flecked portepée knot. Probably no more than a few hundred examples were required, due to the relatively small size of this Police formation, so it was decided that the design should be such that maximum use could be made of existing naval dagger component parts.

In April 1939, this rare sidearm was withdrawn from service and *Wasserschutzpolizei* officers were thereafter required to adopt the standard Police sword instead, emphasising their attachment to Himmler's forces.

## FIRE BRIGADE

During the early years of the Third Reich, each German provincial authority retained its imperial and Weimar responsibility for running a local fire brigade. In 1938, all of these independent formations were brought together and incorporated into the Order Police under the title *Feuerschutzpolizei*, or Fire Protection Police, which thereafter directed fire fighting and fire prevention across the Reich. The size of any given Fire Protection Police contingent was fixed in accordance with the local population, and in those towns with more than 150,000 residents auxiliary fire brigades known as *Freiwillige Feuerwehr* were established on a voluntary basis to assist the regulars. At the height of wartime air raids on Germany, the unified fire-fighting services numbered almost 2 million men and women, backed up by 700,000 Hitler Youth.

### THE FIRE BRIGADE 1870 DAGGER

In 1870, a dagger was authorised to be carried by officer ranks from *Hauptbrandmeister* upwards in the Prussian Royal Fire Brigade. It was 48 cm in length, with a flaming ball pommel, nickel silver fittings and black leather coverings to the grip and scabbard. The crossguard bore a representation of an old brass fire helmet surmounting crossed axes,

**73**. *A Fire Brigade 1870 dagger, by Eickborn.*

while the quillons took the form of trefoils and the blade was etched with miscellaneous fire-fighting motifs. The weapon was hung from two black leather straps with plain nickel silver buckles, and was worn without a portepée knot. From 1871, its use was extended to other fire brigades across the new German Empire.

An acanthus leaf pommel was adopted in 1920. Otherwise, the imperial dagger continued to be sported unaltered by Fire Brigade officers throughout the Weimar Republic era and well into the Third Reich. Surprisingly, the swastika was not added to the design during 1933–4, when other aspects of Fire Brigade uniform were being Nazified. Use of the dagger was officially terminated on 27 May 1936, when commissioned ranks were instructed to wear plain-hilted sabres instead, and with the creation of the *Feuer-schutzpolizei* in 1938 the Police sword was finally authorised for all Fire Brigade officers.

## THE FIRE BRIGADE 1870 AXE

Ornate and heavy 35 cm ceremonial axes were introduced in 1870 as optional items which could be purchased by lower-grade Prussian Fire Brigade members who were not entitled to dress daggers. A range of different qualities and finishes was made available, depending on what the buyer could afford to pay. Each piece commonly featured complex etchings showing fire helmets, floral designs

**74**. *This Fire Brigade 1870 axe was presented to a fireman in 1943, to mark his completion of thirty years' service. The elaborate etching to the steel head is clearly evident.*

and so on, and was suspended from double leather strap hangers attached to two swivel rings on the solid wooden handle. The latter usually bore an oval brass or silver cartouche upon which the owner's name and unit was engraved. Use of the axe spread to the fire brigades of other German states after 1871.

Once again, the dress axe fell into general disuse after 27 May 1936, when bayonets were authorised for off-duty and parade wear by Fire Brigade NCOs and other ranks. However, limited production continued well into the war, with suitably engraved axes being presented to firemen as retirement gifts or as tokens of appreciation for services rendered to their local communities.

## CUSTOMS SERVICE

Formed in 1936, the Nazi Customs Service, or

*Zollgrenzschutz,* comprised two distinct divisions, the *Landzoll*, or Land Customs, and the much smaller *Wasserzoll*, or Water Customs. Operating under the jurisdiction of the Ministry of Finance, its statutory purpose was to 'guarantee the financial sovereignty of the Reich through border security measures'. All imports entering Germany, whether across land boundaries or via her rivers and northern sea ports, had traditionally been subject to extremely severe duties, which had enabled German industry to grow and prosper during the nineteenth and early twentieth centuries, protected by trade barriers. This policy continued under Hitler, and consequently his Customs Service soon developed into an *SS*-supported paramilitary formation which saw its role not only as a tax-gathering body but also as a frontier defence force, an aspect of its work which was given even greater emphasis by the outbreak of the Second World War. After 1939, *Zollgrenzschutz* personnel acted primarily as border police, apprehending Allied prisoners of war, refugees, saboteurs and others trying to make their way out of or into Germany. Indeed, the frontier security function eventually became so dominant that control of the Customs Service was completely removed from the Ministry of Finance on 1 October 1944 and handed over to the *Gestapo*, yet another example of the all-embracing influence of Himmler and the *SS*.

The uniform worn by the *Landzoll* was very similar to that of the German Army, a fact reflected in the design of Customs Service daggers.

## THE LAND CUSTOMS 1937 DAGGER

On 31 July 1937, Reich Finance Minister Lutz Graf Schwerin von Krosigk introduced a plain-bladed 41 cm dagger for wear by *Landzoll* officers from the rank of *Oberzollsekretär* upwards. Similar in style to its Army counterpart, created two years earlier, it featured a Customs-pattern eagle with upswept wings on the crossguard. The wooden grip and steel scabbard were wrapped in dark green leather, the handle being finished off with twisted silver wire and the scabbard with fittings which were peculiarly diagonal in configuration. All

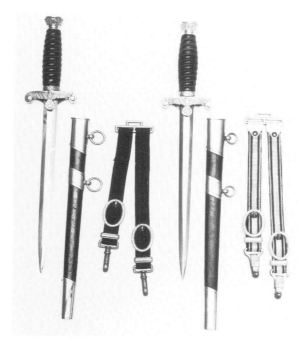

**75**. *Customs Service 1937 daggers*
*Left: Water Customs. Right: Land Customs.*

metal parts were initially die-struck in nickel silver or nickel-plated steel, but from 1938 they were cast in aluminium, which resulted in poorer definition to the design details. The weapon was suspended from a double-strap green velvet hanger with green-edged aluminium braid facings and Army-style attachments, and was worn with a 42 cm aluminium wire portepée knot. The latter had green stripes to the acorn in the case of junior officers.

The principal manufacturer of the *Landzoll* dagger was Carl Eickhorn. Production was discontinued in 1942.

## THE WATER CUSTOMS 1937 DAGGER

*Wasserzoll* officers of the rank of *Maschinenbetriebsleiter* and above, who wore Navy-pattern uniforms, were also authorised to use their own distinctive version of the Customs dagger in July 1937. It was identical in overall design to that of their land-based colleagues, but was distinguished by dark blue leather wrappings and gold-plated bronze fittings. The dagger was hung from black watered silk and velvet straps with gilt buckles, and was finished off with a 42 cm gold portepée knot. It is noteworthy that the quality of detail on the metalwork of *Wasserzoll* daggers was far superior to that of their post-1938 *Landzoll* equivalents, due to the fact that aluminium could not readily be fire-gilded so was never used in the construction of *Wasserzoll* pieces. Die-struck bronze continued to be utilised throughout the production run, which, in any event, was a very small one.

## POSTAL PROTECTION CORPS

Formed in 1934, the *Postschutz*, or Postal Protection Corps, had the responsibility of guarding and maintaining security at all post offices and other postal establishments, and of protecting the mail, telephone and telegraph services throughout the Reich. Prior to 1942 the 4,500-strong *Post-schutz* was under the control of the Postmaster-General, *NSKK-Obergruppenführer* Dr Wilhelm Ohnesorge, but in March of that year, upon the personal directions of Hitler, it was completely incorporated into the *Allgemeine-SS* and redesignated the *SS-Postschutz*. This recognised not only the increasing importance of secure communications during wartime, but also the fact that the majority of *Postschutz* personnel were recruited from Post Office employees who held dual part-time membership of the *SS*. They were almost exclusively older or infirm men, unfit for military service.

The *Postschutz* kept close liaison with an affiliated formation, the *SS-Funkschutz*, or Radio Guard, which policed official radio stations, raided illicit ones and detected illegal listening to foreign broadcasts. Once again, membership was drawn primarily from the *Allgemeine-SS* Reserve.

### THE POSTAL PROTECTION CORPS 1939 DAGGER

On 1 February 1939, Dr Ohnesorge authorised a dagger for wear by *Postschutz* officers of the rank of *Zugführer* and above. Various companies were

76. *The Postal Protection Corps 1939 dagger was an economy piece constructed from parts originally intended for other patterns of edged weapon. The grip, pommel and reworked crossguard were those of an RLB knife; the scabbard that of an RLB 1938 dagger; the crossguard swastika that of an NSFK knife; and the hanging chains those of a Luftwaffe 1935 dagger. Even the grip eagle was adapted from an existing organisational stick-pin badge, and was surface-mounted.*

invited to submit tenders for its design and production, including the firm of Paul Weyersberg & Co., principal manufacturer of the recently revised series of Air Raid Protection League (*RLB*) daggers. In view of the fact that only a very small number of

*Postschutz* pieces were initially required, since there were fewer than 300 qualifying officers at that time, Weyersberg simply reworked some readily available *RLB* dagger parts and added a *Luftwaffe* chain hanger, a swastika badge from the *NSFK* knife and a distinctive grip eagle to produce a very cheaply priced prototype which met with the approval of the postal authorities.

The plain-bladed 39 cm *Postschutz* dagger had a black grip and scabbard with nickel silver fittings. The handle bore the organisation's emblem of an eagle surmounting lightning bolts, while the cross-guard took the form of a stylised eagle's head with a black enamel swastika on the quillon block. The weapon was suspended from a plain double chain hanger made up of nickel silver rings, and was worn with a 42 cm aluminium portepée knot inter-woven with orange threads.

*Postschutz* daggers remained Post Office proper-ty and could not be bought on the open market. Each example was issued centrally, with small quantities being stored at main postal establish-ments. Officers requiring daggers while on duty were therefore obliged to sign them out from local stocks and return them again after use. As a control measure in this respect, every piece bore an accountability serial number stamped under the crossguard, together with the 'DRP' ownership mark of the *Deutsche Reichspost*.

# RAILWAY SERVICE

During the First World War Germany operated the *Eisenbahn*, the most extensive national railway network in Europe. It provided an unprecedented capability for the rapid movement of enormous numbers of troops and their equipment, and enabled the Kaiser to maintain a war on two fronts for three years. As a result of the Versailles settle-ment, the *Eisenbahn* was made a subject of war reparations, with much of its rolling stock being seized by the French. The centralised German railway system was broken up by the victorious Allies, and responsibility for running domestic train services was devolved to individual state governments, including those of Bavaria, Prussia, Saxony and Württemberg.

After assuming power, the Nazis determined to restore the strategic nature of the railways, and in 1935 the various local railway authorities were reunited to form the *Deutsche Reichsbahn*, a nationally owned public service corporation which subsequently enjoyed a considerable measure of financial, administrative and operational autonomy. It duly expanded to include the railway systems of Austria and Czechoslovakia, and by 1939 covered 35,000 miles of track operated by 800,000 employ-ees. With the outbreak of the Second World War, the *Reichsbahn* again played an ever more important role in transporting military personnel and materi-als throughout the occupied territories. By the time it reached its manpower peak of 1,400,000 staff in 1942, the railway service had in effect become a crucial auxiliary branch of the *Wehrmacht*.

The *Bahnschutz*, or Railway Guard, was main-tained by the *Reichsbahn* to protect railway property and prevent theft and sabotage. It worked closely with the *Bahnpolizei*, or Railway Police, an affiliated group which operated a patrol service to keep order and discipline among railway employ-ees and train passengers, and with the *Reichsbahn-Wasserschutzpolizei*, or Rail Waterways Protection Police, which patrolled railway facilities associated with harbours, canals and inland waterways. These three formations were recruited primarily from *Deutsche Reichsbahn* personnel who held part-time membership of the *Allgemeine-SS* or *SA*, and their staff were armed with rifles and machine-guns. The overlapping functions and jurisdictions of the *Bahnschutz*, *Bahnpolizei* and *Reichsbahn Wasser-schutzpolizei* resulted in their being amalgamated in 1941 under the all-embracing title of *Bahn-schutzpolizei*. The new body was charged by the *Gestapo* with the additional specific responsibility of looking out for 'wanted persons' on trains and at railway stations, with a view to countering wartime espionage. Not surprisingly, the *Bahnschutzpolizei* was soon removed from the auspices of the *Reichsbahn* to come under the umbrella of Himmler's National Police Force, and from 1944 it became known as the *SS-Bahnschutz*.

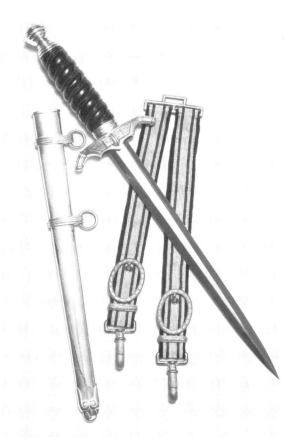

**78**. *The squirrel (or* eichhörnchen*) trademark of Carl Eickhorn, in its 1938-41 version designed by Paul Casberg, appeared on the blades of all* Bahnschutz *daggers.*

**77**. *The Railway Guard 1938 dagger.*

## THE RAILWAY GUARD
### 1938 DAGGER

In April 1938 a dagger was created for wear by Railway Guard officers of the rank of *Oberzug-führer* and above, as part of a newly introduced uniform. Although referred to as the *Bahnschutz-führerdolch*, it was also worn by relevant officials of the *Bahnpolizei*. The 40 cm plain-bladed dagger had polished aluminium fittings, a deep purple (almost black) grip and a spherical pommel embossed with a raised sunwheel swastika. The crossguard featured the winged railway wheel insignia of the *Bahnschutz*, while the smooth scabbard bore ribbed suspension bands and an ornate scrolled pattern to the tip. The weapon was worn from a double-strap fabric hanger comprising aluminium braid facings with black edge stripes sewn to purple velvet backings, and it sported a 42 cm aluminium wire portepée knot interwoven with purple threads.

The *Bahnschutz* dagger was designed by Paul Casberg for exclusive manufacture by Carl Eickhorn. Production ceased in 1941, shortly before the formation of the *Bahnschutzpolizei*.

## THE RAIL WATERWAYS PROTECTION
### POLICE 1938 DAGGER

When the *Bahnschutz* and *Wasserschutzpolizei* sidearms were authorised independently within days of each other in the spring of 1938, the railway authorities felt it would be prudent to introduce a further distinctive dagger for officers of the *Reichsbahn-Wasserschutzpolizei*. Design and manufacturing rights were again vested solely in the Eickhorn firm. Due to the extremely limited number of daggers required for this small unit, Eickhorn decided to simply cobble together parts from existing production patterns to create a new weapon. The design subsequently chosen combined the pommel, grip and scabbard bands of the *Bahnschutz* dagger with the crossguard and scabbard of the Army dagger. All metal parts were gilded, to match the Rail Waterways Protection Police uniform. The hybrid sidearm was suspended from an equally cost-effective and utilitarian double-strap black leather hanger, and was worn with a 42 cm gold bullion portepée knot.

Manufacture was discontinued in 1941, when the *Reichsbahn-Wasserschutzpolizei* ceased to exist as a separate entity.

## THE RAILWAY SERVICE 1941 DAGGER

In February 1941, a new uniform featuring an arm eagle and unit cuff title was introduced for train drivers, communications officials, surveyors and all other general personnel of the *Deutsche Reichsbahn*, in recognition of their increasingly paramilitary status. The Railway Service did not use a rank system based on that of the *Wehrmacht*. Instead, its employees were divided into four main worker classifications and subdivided into twenty-three pay groups containing eighty-three different grades of official. For the first time, those in the top two worker classifications were now accorded the distinction of wearing a dress dagger with their new uniform.

Once again, the rail authorities turned to Paul Casberg to produce a suitable design. He proposed a weapon similar in style to the *Luftwaffe* 1937 dagger. The plastic grip and steel scabbard were dark blue in colour, with all other fittings being gilded. The pommel featured a swastika within an oakleaf wreath, while the crossguard bore the *Reichsbahn* winged wheel insignia. The scabbard suspension bands and chape were gilt, and each sported an oakleaf motif. The dagger was to be suspended from gold brocade hangers with blue

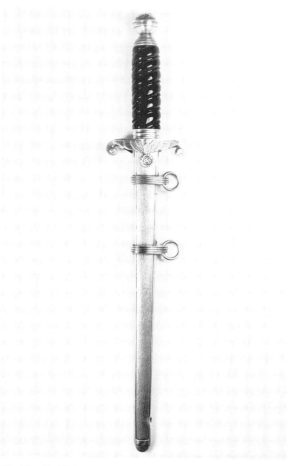

79. *The Rail Waterways Protection Police 1938 dagger was a 'parts' creation, utilising an Army crossguard and scabbard body married to fittings from the* Bahnschutzführerdolch. *The gilded finish, however, gave it a very distinctive appearance.*

velvet backs, and would be complemented by a gold portepée knot with blue flecks. Casberg's design was duly accepted. At least one prototype was manufactured by Eickhorn and was depicted in a photograph showing the new *Reichsbahn* uniform being worn, published in the March 1941 edition of the trade journal *Uniformen-Markt*.

However, no further examples are known to have been made. Ceremonial blade production in general was then being discontinued for the duration of the war, because of the shortage of iron, steel and

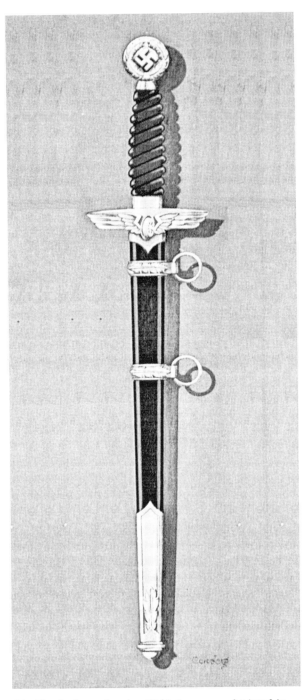

other strategic materials, and it is likely that the *Reichsbahn-Dolch* was put on hold for reintroduction in peacetime. Indeed, by September 1941 even the short-lived railway cuff titles were being withdrawn for reasons of economy, with unit designations incorporated into revised one-piece arm eagle badges instead.

## LABOUR SERVICE

In 1931, Hitler charged ex-Army officer Konstantin Hierl with the setting up of a Nazi Party voluntary labour service, the *NS-Freiwillige Arbeitsdienst*, or *FAD*, in an effort to ease the unemployment then rife amongst *NSDAP* members across Germany. Hierl's conception of his new organisation went far beyond the need to counter joblessness, however. He believed that universal manual labour could provide an ideal means of moulding the character of the young and of breaking down social and class barriers. With the Nazi assumption of power two years later, Hierl was confirmed as *Reichsarbeitsführer* and appointed Secretary of State for Labour. The *FAD* now ceased to be a party formation, being elevated to the status of a national authority and renamed the Reich Labour Service, or *RAD*. From 26 June 1935, six months' service in the *Reichsarbeitsdienst* was made compulsory for all males between the ages of seventeen and twenty-five, and, once the machinery of military conscription got under way, this obligation was normally fulfilled prior to induction into the armed forces. By 1939, *RAD* membership stood at 360,000 men, who were organised and administered by a relatively small cadre of permanent full-time staff.

In peacetime the *RAD* undertook land reclamation, drainage projects and soil conservation work, and was instrumental in the construction of the *autobahn* system. With the outbreak of the Second World War it became an important auxiliary formation of the *Wehrmacht*, constructing fortifications, roads and airfields. Its personnel also repaired railways, laid minefields, manned defensive bunkers, built air-raid shelters and

**80**. *The Railway Service 1941 dagger, as depicted in the original design drawing by Paul Casberg.*

**81.** *An* RAD *1934 hewer, with its characteristic heavy-duty brown leather hanger.*

**82**. *An* RAD-Obertruppführer *from the 10th* Abteilung, *301st* Gruppe, *posing with his hewer in 1940.*

carried out a range of other duties. By 1943 the *RAD* was fully armed and completely militarised, and was operating throughout the occupied territories. A female youth section was also very active on the home front, servicing factories and munitions plants, the agricultural industry, war-relief agencies and the public transportation system.

The vital contribution made by the *RAD* to the Nazi war effort can be gauged by the fact that on 24 February 1945, his seventieth birthday, Hierl became only the second living recipient of the German Order, the Third Reich's highest honour.

## THE *RAD* 1934 HEWER

At the beginning of 1934, a heavy 40 cm hewing knife, or *Haumesser*, was authorised by Hierl for wear by all career officers and NCOs of the *RAD* on

formal or ceremonial occasions. Created by Paul Casberg and based on an old German woodcutting hatchet, it featured staghorn grips and a massive scimitar-shaped blade on which was etched the Labour Service motto *Arbeit Adelt* ('Work Enobles') and the *RAD*'s distinctive triangular proof mark. The stove-enamelled blackened steel scabbard bore swirling Germanic ornamentation embossed into the nickel silver or nickel-plated steel locket, while the chape sported the organisation's emblem of a swastika-charged spade head over two ears of barley. The sidearm was suspended from a large brown leather hanger and was worn without a portepée knot.

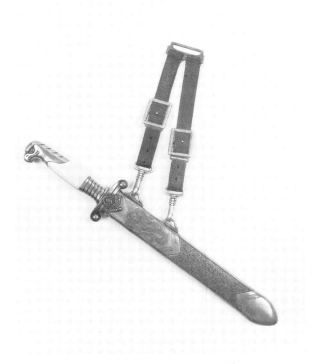

**84**. *An* RAD *leader's 1937 hewer, by Eickhorn*

**83**. *The triangular* RAD *approval mark and the Ges.Gesch. patent pending designation are clearly evident on the blade of this 1934 hewer produced by Eduard Wüsthof.*

The *Haumesser*'s introduction was heralded by the Press announcement:

> The hewer of the Reich Labour Service is designed in conformance with the *RAD*'s non-combatant activities. It is not only a dress uniform accoutrement but also a practical tool for the cultivation of our soil. It is therefore symbolic of the objectives of the Reich Labour Service, namely to secure new land for our people and a new people for our land.

The 1934 hewer was not made available for private purchase. Stocks were held on a local battalion basis, for distribution and return as required, and each piece was normally marked with a unit abbreviation and an accountability serial number. From

1937, issue to officers was discontinued and the 1934 hewer was thereafter reserved solely for cadre NCOs ranking between *Truppführer* and *Unterfeldmeister*. Production ceased during the middle of 1943.

## THE *RAD* LEADER'S 1937 HEWER

In December 1937, officers of the rank of *Feldmeister* and above serving on the permanent staff of the *RAD* were rewarded with their own version of the hewer, again devised by Casberg, which was significantly finer and more dagger-like in appearance than the 1934 pattern. The handle took the form of an eagle's head pommel surmounting a white, orange or ivory-coloured plastic grip, while the crossguard sported the ubiquitous *RAD* spade insignia. The scimitar blade was elongated in shape

85. RAD-Feldmeister *Werner Schrammkrug, attached to the 5th Abteilung, 156th* Gruppe, *wearing his leader's hewer on the occasion of his marriage to Gertraud, 12 May 1940.*

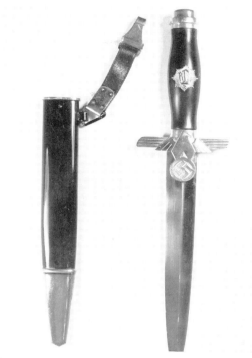

86. *An* RLB *1936 knife. Of particular note is the unique triangular fitting at the throat end of the scabbard, to which was riveted the hanging strap.*

and the scabbard mirrored the general ornamentation of its 1934 predecessor. All metal fittings were silverplated or polished aluminium. The *Führerhaumesser* was suspended from a double-strap brown leather hanger with rectangular white metal buckles, and was worn without a portepée knot. It could be purchased and retained by individual officers, a few of whom had it engraved with their names or initials. Manufacture was discontinued in 1943.

It is interesting to note that Hierl was presented by his senior officers with a unique version of the 1934 hewer, having gold-plated fittings, a Damascus steel blade and double leather strap hangers, which he continued to wear long after the introduction of the standard officer's pattern.

# AIR RAID PROTECTION LEAGUE

The Air Raid Protection League, or *Reichsluftschutzbund* (*RLB*), was founded by Hermann Göring on 28 April 1933. Operating under the auspices of the Air Ministry, its small cadre of salaried uniformed staff supervised civil defence training throughout Germany, arranging practical demonstrations, lectures and film shows on a local basis. All members of the public were encouraged to volunteer for service with the organisation, and by 1939 over 15 million people had done so. Their duties were normally restricted to the running of air-raid shelters, which were constructed on a street-by-street basis across the country. Volunteers were not provided with a uniform, and were obliged to purchase their own special steel helmets,

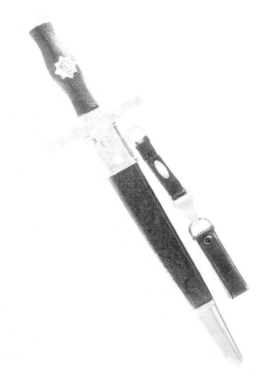

**87**. *An* RLB *dagger, with a belt loop attached to its regulation hanging strap.*

which were generally worn with suitable civilian clothing.

During the Second World War, *RLB* responsibilities were extended to include firefighting, search and rescue, decontamination and building demolition. In common with other home front forces, the *Luftschutz* came to be increasingly influenced by the *SS* and Police, with *RLB* recruiting being carried out almost exclusively at local police stations.

## The *RLB* 1936 Knife

A 36 cm knife was authorised during the summer of 1936 for dress wear by selected full-time NCOs and subordinate personnel of the *RLB*, ranking between *Obertruppmeister* and *Truppmann*. Manufactured principally by the firm of Paul

Weyersberg, the weapon featured a plain 22 cm blade and a black wooden handle. The latter bore the first pattern *Luftschutz* insignia comprising an eight-pointed sunburst surmounted by the initials '*RLB*' over a swastika. The knife's crossguard took the form of a stylised short-winged eagle and swastika which, like the dome-shaped pommel, was nickel-plated. The black stove-enamelled scabbard had a nickel-plated chape, and was suspended from a leather strap permanently affixed to a unique black-painted triangular fitting at its unadorned throat end. No portepée knot was used with the knife. The *RLB-Messer* was conferred by special order only upon those deemed to be worthy recipients, with qualification criteria based on length of service and meritorious conduct. Consequently, it was not available commercially for private purchase.

## The *RLB* 1936 Dagger

At the same time as the introduction of the *Luftschutz* knife, a 39 cm dagger was created for bestowal upon selected salaried *RLB* officers between the ranks of *Führer* and *Präsident*. It was similar in general appearance to the knife, but was produced in a much finer quality and was distinguished by a number of characteristic features. Most notably, both the grip and scabbard were wrapped in blue-black Morocco leather. The pommel was flattened in configuration, the wings of the crossguard eagle were extended, its swastika was silhouetted and the blade was drawn out to a more elegant 25 cm length. Finally, the scabbard had the addition of a silver-plated locket to which was attached a standard nickel suspension ring for the carrying strap. Like the knife, the *RLB-Führerdolch* was worn without a portepée knot and was conferred only upon deserving individuals, usually *Amtsträger*, or office holders who bore particular responsibilities.

## The *RLB* 1938 Knife

The sunburst insignia of the *Luftschutz* organisation was redesigned during the summer of 1938 to

delete the initials 'RLB'. The revised badge comprised simply a large black swastika set upon the corporate eight-pointed sunburst. From 29 August 1938, the knives issued to *RLB* NCOs and subordinates included this new emblem on their handles. Otherwise, they were identical to their 1936-pattern predecessors.

### THE *RLB* 1938 DAGGER

In August 1938, the form of the *RLB* officer's dagger was altered to feature not only the updated *Luftschutz* grip insignia but also a detachable double-strap black leather hanger with oval white metal buckles bearing oakleaves. To facilitate the wear of this new accoutrement, a plain silver-plated mount with suspension ring was added to the centre of the scabbard. The rest of the dagger specification remained unchanged from that of the 1936 version.

Production of *RLB* daggers and knives ceased in 1942.

# TECHNICAL EMERGENCY CORPS

The *Technische Nothilfe*, or *Teno* (Technical Emergency Corps), was founded in September 1919 as a government strike-breaking organisation and later used as a resource in times of natural or industrial disaster. It was incorporated into the *Ordnungspolizei* in 1937, although it retained its own series of uniforms, and was set the task of dealing with breakdowns in public services and utilities such as gas, water and electricity, particularly after air raids. From 1939, another more remote purpose of *Teno* was to prepare for anticipated revolutionary conditions in the 'Theatre of War Inner Germany', as Himmler liked to call the home front where the majority of his forces operated. In addition to this varied domestic work, units known as *Teno Kommandos* laboured with the *Wehrmacht* and *RAD* on front-line construction and repair. Membership of the Technical Emergency Corps peaked at 100,000 in 1943. However, the vast majority of *Teno* personnel were unpaid

**88**. *The* RLB *1938 knife.*

**89**. *An* RLB *1938 dagger, complete with leather hanging straps.*

volunteers, normally older or disabled ex-servicemen possessing specialist skills as engineers, plumbers, electricians and so on. Only a small number of full-time salaried officers and NCOs were employed to administer the corps and control its staff.

## THE *TENO* 1938 HEWER

A massive 40 cm hewing knife, similar in appearance to that of the *RAD*, and again created by Paul

**90**. *The* Teno *1938 dagger (left) and hewer (right). The scabbard of the latter has its unique pattern of suspension frog attached.*

Casberg, was introduced on 30 November 1938 for dress wear by selected full-time *Teno* NCOs. Its oxidised nickel-plated steel hilt sported a cogwheel emblem on the pommel, white or orange plastic grip plates and the *Teno* eagle on the crossguard. The scimitar-shaped blade was contained in a black stove-enamelled steel scabbard with nickel fittings, suspended from a black leather frog. A range of knots in different colours was made available for use with the hewer, depending on the primary duty or trade of the wearer. The most common type had a silver acorn with a purple top and orange stem, indicating membership of the reserve. A stem in red denoted expertise in post-air-raid work, while blue related to technical service and green to more general duties. These colours corresponded to the *Waffenfarbe* piping used on *Teno* uniforms.

Carl Eickhorn secured exclusive rights to produce the *Haumesser*, but the *Teno* High Command retained the copyright to its actual design, having commissioned Casberg directly. Each piece was therefore etched on the blade ricasso not only with Eickhorn's ubiquitous squirrel trademark but also with the *Teno* eagle surmounting the legally protective *Ges. gesch.* ('Patent pending') designation. Moreover, the interior sides of the grip plates were embossed with squirrel logos which were visible only upon disassembly. *Teno* hewers could not be obtained on the open market and were centrally issued on a unit by unit basis, for return to local stores after use. Consequently, each example was stamped on its blade and scabbard with matching accountability serial numbers as well. Manufacture was discontinued in 1943.

## THE *TENO* 1938 DAGGER

At the same time as the introduction of the *Haumesser*, an Army-style dagger was authorised for all regular *Teno* officers of the rank of *Kameradschaftsführer* and above. Its pommel and crossguard bore cogwheel and eagle devices identical to those on the hewer, while the grip was constructed of yellow or ivory-coloured plastic with horizontal grooves. The 27 cm plain steel blade was held in a pebbled scabbard with ribbed

**91**. *The ricasso of the* Teno *1938 dagger, showing the organisation's eagle surmounting the protective* Ges.Gesch. *designation and the Eickhorn squirrel logo.*

suspension bands and an ornamental tip. All metal fittings were in silver-plated aluminium, oxidised to give a tarnished appearance. Production and design rights were once more held jointly by Eickhorn and *Teno*, with appropriate markings on the ricasso. Blades and scabbards likewise bore corresponding serial numbers, as the daggers were again issued only via unit stores and could not be bought commercially.

Uniquely, the *Teno-Führerdolch* was provided with two different styles of double-strap hanger, one for parade wear and one for more general daily use. The former featured black velvet backings faced with silver stripes, while the latter took the form of simple black or brown leather straps. Both types bore cogwheel-shaped buckles. The dagger

was worn with a *Luftwaffe*-pattern 23 cm aluminium portepée knot. Manufacture ceased in 1943.

# RED CROSS

During the early years of the Third Reich, the German branch of the International Red Cross Society, with its headquarters in Switzerland, was the main social welfare organisation tending to the needs of the German people in times of hardship. On 9 December 1937, Hitler conferred a new legal status on the *Deutsches Rotes Kreuz* by recognising it as a national corporation with a degree of independence from Geneva. The *DRK* thereafter expanded considerably in size and responsibility, with two distinct branches developing. One remained primarily active in the fields of medicine, nursing and first aid, while the other concentrated on the provision of a variety of charitable and social-work services ranging from child maintenance to the feeding and accommodation of the elderly and homeless. During the Second World War, the German Red Cross was heavily involved on both the home and battle fronts, while its international dimension gave it a unique authority so far as the tracing and monitoring of prisoners of war were concerned. The majority of Red Cross personnel were volunteers, administered by a cadre of permanent officers and NCOs under the leadership of *DRK-Präsident* Prof. Dr Ernst-Robert Grawitz. Perhaps not surprisingly, Grawitz was also an *SS-Obergruppenführer* and Chief *SS* and Police Medical Officer. Even the Red Cross could not avoid becoming entangled in the web-like structures of Himmler's ever-expanding *SS* empire.

### THE *DRK* 1938 HEWER

In February 1938, a 40 cm hewer was authorised for daily wear by full-time Red Cross NCOs and subordinates ranking from *Helfer* to *Haupthelfer*. Produced principally by Robert Klaas and P.D. Lüneschloss, it featured a flared pommel, black bakelite grip plates and an oval crossguard cartouche bearing the *DRK* emblem of an eagle with a swastika on its chest, holding the International

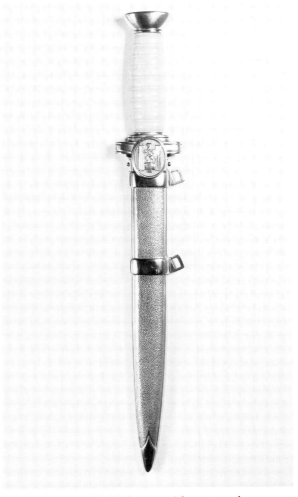

**92**. *The Red Cross 1938 hewer was a massive utility sidearm which could be put to practical use when preparing splints or removing plaster casts.*

**93**. *A Red Cross 1938 dagger, with rectangular suspension ramps.*

Red Cross badge in its talons. The broad blade had a sawtooth edge for use in preparing splints and removing plaster casts, while the point was squared off and blunted to conform with the restrictions of the Geneva Convention which prohibited the carrying of offensive weapons by medical or first-aid personnel. The *DRK-Haumesser* was held in a black enamelled scabbard with nickel-plated locket

and chape, suspended from a black leather frog, and was worn without a portepée knot. It could not be obtained on the open market, and was issued from local unit stores only when required. Most pieces were therefore stamped on the reverse langet with appropriate accountability markings and serial numbers. Manufacture was discontinued in 1940 for the duration of the war.

### The *DRK* 1938 Dagger

A 37 cm dagger was also introduced in February 1938 for dress wear by regular Red Cross officers of the rank of *Wachtführer* and above. Its nickel-plated hilt comprised a white or yellow plastic grip complemented by a pommel and crossguard similar in style to those of the hewer. The blade was pointed, which still complied with the Geneva Convention as the dagger was not intended to be carried in the field, and was contained within a pebbled nickel scabbard with two circular or rectangular suspension ramps. The shape of these ramps varied between manufacturers, the main firms involved being Robert Klaas and Carl Eickhorn. Once again, the *DRK-Führerdolch* was centrally issued and could not be bought commercially. No portepée knot was authorised by regulations, although it was fairly common for Red Cross officers with military backgrounds or assigned to the armed forces to sport a 42 cm aluminium knot wrapped round the lower end of the grip.

Two entirely different sets of double-strap hangers were made available for the dagger, depending upon the duties of the wearer. Those staff involved in medical and first-aid work utilised tan-brown velvet straps faced with red-bordered aluminium stripes, finished off with smooth oval buckles. Social welfare officers, on the other hand, wore grey velvet hangers featuring blue-bordered aluminium facings and pebbled rectangular buckles. Production ceased in 1940, although the *DRK* dagger continued to be carried on dress and ceremonial occasions well into the war.

## FORESTRY SERVICE

Over 800,000 people were employed in Germany's woodworking industries during the 1930s, and forests were recognised as being an important natural resource. Paper manufacturers, printers, dye-makers and a chain of connecting interests all relied on the efficient running of the vast tree plantations which stretched across Bavaria, Prussia and the other rural regions of the Reich. In 1934, the Nazis centralised forest

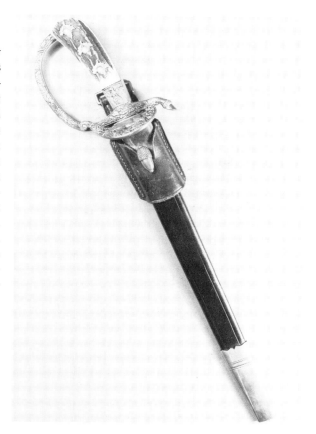

**94**. *A Forestry Service 1938 cutlass by F.W. Höller, with staghorn grips for junior ranks. Uncommonly, this example has an eagle and swastika emblem cast above the crossguard.*

administration under the *Reichsforstdienst*, or National Forestry Service, headed by Hermann Göring, whose officials were made responsible for conserving, developing and generally maintaining all wooded areas, whether in public or private ownership. They worked in close co-operation with their colleagues in the Army and *Luftwaffe* forestry services, who managed woodland sited on military training grounds and other *Wehrmacht* estates. With the outbreak of the Second World War and the acquisition of large, fertile territories in the East, lumbering in Poland and Russia greatly enlarged the scope of *Reichsforstdienst* activities. *Ad hoc* economic operations units, or *Wirtschafts-kommandos*, roamed the countryside co-ordinating local entrepreneurial projects, and between 1941 and 1944 the Forestry Service assisted the *SS* in exploiting the wealth, resources and population of the conquered East on a massive scale.

### THE FORESTRY SERVICE 1938 CUTLASS

Cutlasses, or *Hirschfänger*, resembling short swords, had been worn by all German foresters since the eighteenth century. During the early years of the Third Reich, there existed no regulation pattern and dozens of manufacturers offered large numbers of extreme variations for private purchase. Some degree of standardisation was introduced for *Reichsforstdienst* officials in June 1938, although different lengths, qualities and styles of ornamentation were still permitted within set limits. The main common denominators prescribed by the National Forestry Service were knuckle bows, shell guards, gilded metal fittings and blades etched with woodland scenes. The handle of every regulation cutlass featured three sets of acorns with oakleaves, while the grip plates were in brown staghorn for ranks below *Förster* and ivory or white celluloid for officers. Each sidearm was carried in a brown or black leather scabbard with a brass locket and chape, suspended from a frog.

Enlisted men and NCOs from *Anwärter* to *Unterförster* wore a plain green portepée knot, while those officials graded *Förster* and higher used a silver knot with green stripes. The most senior officers, ranking *Oberlandförster* and above, sported gold knots with green stripes. Commissioned officers in the Army and *Luftwaffe* forestry services were allowed to wear their standard dagger knots with the *Hirschfänger*.

The Forestry Service cutlass was one of the few Nazi edged weapons which did not routinely feature the swastika as part of their design. Manufacture ceased in 1942.

# HUNTING ASSOCIATION

In 1934, in an effort to regulate the culling of deer, boar and other wild game which roamed freely throughout the German countryside, *Reichsjägermeister* Göring subjected the numerous existing hunting fraternities to the authority of a new National Hunting Association, the *Deutsche Jägerschaft*, or *DJ*. Its main purpose was to further the cause of livestock conservation by limiting the extent of shooting during the hunting season and protecting endangered animals during their breeding periods. To these ends, Hunting Association functionaries worked in close co-operation with personnel of the *Reichsforstdienst*. By 1935, the *DJ* was firmly established as a disciplined and fully uniformed organisation.

**95**. *A Reichsforstdienst officer wearing the enlisted man's/NCO's pattern forestry cutlass, adorned by a knot from an Army officer's sword. Such irregular 'personalised' combinations were not uncommon among members of this organisation.*

## THE HUNTING ASSOCIATION 1936 CUTLASS

Professional and amateur German hunters, like their forester colleagues, had used a wide selection of practical working knives since the eighteenth century. On 22 March 1936, the *DJ* introduced a series of regulation 49 cm *Hirschfänger* for wear by its officials and ordinary members as dress weapons rather than tools of the trade. They were similar in style to the cutlasses of the Forestry Service, but were without knuckle bows. Once again, many slight variants were permitted. The metal fittings were generally silvered, with the quillons taking the form of deer hooves or, less commonly, leaves. The ubiquitous staghorn grips, which could be straight or curved, sported the National Hunting Association emblem of a deer skull above the initials '*DJ*'. Between the skull's antlers, a swastika was surmounted upon the cross of St Hubert, patron saint of hunters. Blades were generally etched with hunting motifs and were made available in an alternative non-regulation 34 cm length. Each cutlass was held in a green leather scabbard with nickel fittings, suspended from a frog. On 29 January 1937, a silver portepée knot with gold stripes was authorised for use by professional hunters only.

**97**. *Members of the* Deutsche Jägerschaft *sounding a fanfare at Göring's hunting lodge in 1937. All wear the regulation* DJ *cutlass.*

**96**. *Deutsche Jägerschaft 1936 cutlass, by Alcoso, sporting the silver and gold knot worn by professional hunters.*

Production of the *DJ-Hirschfänger* was discontinued in 1942.

# RIFLE ASSOCIATION

Hundreds of provincial civilian rifle clubs were well established throughout Germany long before the advent of the Third Reich. Shooting contests were traditionally gala events at which competitors took their marksmanship very seriously, and various colourful club uniforms, medals and edged weapons were worn proudly during these proceedings. Hitler considered this particular sport to have special potential in the field of pre-military training, and in 1934 he brought the German Rifle Association, or *Deutscher Schützen Verband* (*D.Sch.V.*), to which all local shooting clubs were required to be

affiliated, under the direct control of the *NSDAP*. By 1936, a prescribed pattern of national *D.Sch.V.* uniform had been agreed, although general wear by all association members, who were obliged to pay for the outfit, was not achieved until 1937–8.

**98.** *This* Hirschfänger *is typical of those carried by rifle club members prior to February 1939. Clearly derived from the hunting cutlass, its shell guard is embellished with a target, crossed rifles and oakleaves, surmounted by an alpine hat. The knot is dark green in colour.*

### THE RIFLE ASSOCIATION 1939 CUTLASS

In February 1939, a regulation 54 cm cutlass was authorised for optional wear with the new *D.Sch.V.* uniform. Similar in appearance to the *Deutsche Jägerschaft Hirschfänger*, it had nickel-plated fittings with a white celluloid grip bearing a gilded crossed rifles badge. The curved quillons featured acorns at each end, while the shell guard sported the *D.Sch.V.* insignia of a black enamel eagle, a target and two swastikas. The weapon's blade was etched with shooting motifs, which could vary according to the taste and choice of the purchaser. Shorter non-regulation blades were also made available by some commercial companies. Each cutlass was contained in a black leather scabbard with nickel fittings, and was suspended from a black leather frog to which was attached a plain green portepée knot.

Manufacture of the *D.Sch.V. Hirschfänger* ceased in 1940, by which time the majority of marksmen were on active service with the *Wehrmacht*.

While the authorisation of dress daggers by the Nazi regime was wide-ranging, it was not exhaustive. A number of influential uniformed

*99. Rifle Association 1939 cutlass by Eickhorn, complete with its black leather frog.*

junior *Politischen Leiter* were sworn in at a single ceremony in Munich on 20 April 1938, and such events took place annually. However, it is noteworthy that local *Ortsgruppenleiter* and more senior Nazi party officials serving in the *Kreis, Gau* and *Reich* levels of leadership were instead permitted to carry 7.65 mm Walther PPK pistols as their *Ehrenwaffen* or 'weapons of honour', held in suitably ornate holsters.

Adolf Hitler, as supreme commander of the *SA, SS, NSKK, HJ* and so on, to say nothing of the armed forces, was technically entitled to sport many different daggers by virtue of his personal authority, but chose not to do so. This may have been due to his inherent dislike of the overtly theatrical, demonstrated by the fact that he wore the plainest of uniforms and only very few of the many decorations for which he undoubtedly qualified. The sole occasion on which Hitler was seen carrying a dagger was during his state visit to Rome in May 1938, when Mussolini awarded him honorary rank in the *MVSN*, the Italian Fascist militia, and presented him with a special ivory-handled version of the *MVSN podesta*. For the sake of politeness, Hitler took pains to be photographed adorned with this edged weapon throughout the five-day tour, after which it disappeared from view upon his return to Germany and was never seen again.

Throughout the life of the Third Reich, several Solingen cutlers submitted designs for new daggers which, for a variety of reasons, were never accepted by the Government. Carl Eickhorn, for instance, had Paul Casberg work up sketches for proposed Police officers' and mining officials' daggers in 1936 and 1938 respectively. Another Eickhorn designer by the name of Köhler illustrated a suggested dagger for the *Weer Afdeelingen*, the Dutch Nazi equivalent of the *SA*, in 1940. While these projects never progressed, the Solingen edged weapons factories continued to expand their sales bases in other ways, notably by winning contracts to supply dress daggers to the armed forces of Bulgaria, Finland, Hungary, Romania and other friendly powers. The turn of the tide of war from 1942, however, necessitated an end to the production of such ornate but basically impractical items

paramilitary formations such as the *DAF-Werkscharen*, whose members acted as factory shop stewards, and the *Organisation Todt*, which co-ordinated strategically crucial military and civil engineering operations, were never considered for the granting of dress daggers. Other groups like the Criminal Justice and Prison Services were allocated more traditional swords. The most striking omission of all was the vast Corps of Political Leaders of the *NSDAP* itself, which was allowed no edged weapons whatsoever. Perhaps this was because of the sheer numbers involved – over 600,000 new for reasons of economy. By May 1943, manufacture had all but ceased. The fortunes of the dagger as a symbol of Nazi military might and civic splendour therefore mirrored almost exactly those of the Third Reich itself, peaking in 1940–1 and thereafter declining sharply towards oblivion.

Before leaving the subject of daggers, one feature common to many of them merits brief coverage. For years, collectors were mystified as to why considerable numbers of Nazi daggers exhibit deep dents to the lower end of their scabbards. Various theories for this damage were put forward, including one that it had been inflicted as a sort of 'mark of surrender' by the Allied authorities responsible for gathering in Nazi edged weapons at the end of the war. However, the real explanation is much more straightforward and obvious. Third Reich servicemen did the majority of their regular travelling by train, and swinging daggers were continually being caught in the sliding doors of railway carriages. A dented scabbard is therefore a fairly conclusive sign that the piece concerned was worn with pride during at least one journey on the *Reichsbahn*!

**100**. *Paul Casberg at work in his Berlin studio, c. 1938. The name 'Paul Casberg' was in fact a pseudonym used by this prolific freelance artist, who produced designs not only for Carl Eickhorn but also for various military regalia firms, particularly in Lüdenscheid, and for the textile industry in Spain. He created a whole range of assorted items, from flags and belt buckles to the Göring Wedding Sword and the baton carried by German field marshals. Casberg's studio, sited near Berlin's main radio tower, was bombed out in 1944 and he is believed to have committed suicide during the fall of the Reich capital in April the following year.*

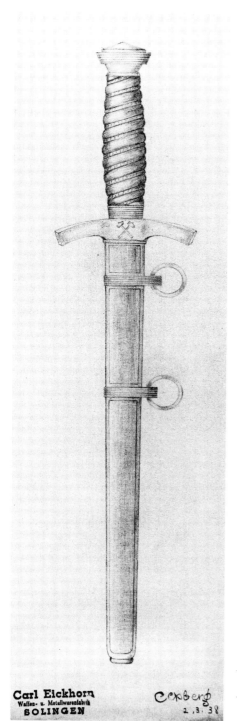

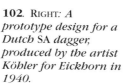

**101**. LEFT: *A prototype design for a Bergbau, or mining industry, official's dagger created by Paul Casberg for the Eickhorn firm on 2 March 1938. The crossguard is shaped like two hammer heads, and bears two crossed hammers. This item never progressed beyond the drawing board. Of particular interest are the Eickhorn property stamp at lower left, and Casberg's signature at lower right.*

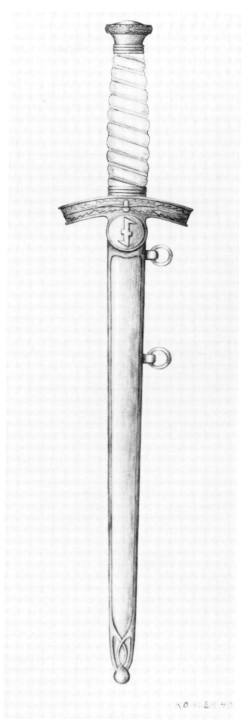

**102**. RIGHT: *A prototype design for a Dutch SA dagger, produced by the artist Köhler for Eickhorn in 1940.*

# PART TWO

# SWORDS

Unlike daggers, regulation swords had a long tradition of widespread use in Germany. They retained their popularity during the Nazi period, being worn as symbols of authority not only by the military but also by members of a variety of uniformed civil organisations, particularly those concerned with the maintenance of law and order. Personnel belonging to strictly political formations were not officially entitled to sport swords, except for those in the *SS* and certain elite units of the *SA*, who could carry them because of their military and Police-related functions.

Third Reich swords were grouped into three distinct categories, depending upon their design characteristics:

- The *Schwert*, or sword proper, featured a straight double-edged blade and a heavy cross-like hilt, and traced its origins to the hacking and cleaving weapons used by medieval warriors.
- The *Degen*, or rapier, originated as a light thrusting weapon during the sixteenth century and was later adopted widely as a court sword, or dress sword. It typically had a straight single-edged blade, a plain grip and a C-shaped knuckle bow.
- The *Säbel*, or sabre, was intended for cutting or slashing and had been popular with cavalry troops since the eighteenth century. It was distinguished by a curved blade, an indented grip and an ornate knuckle bow to the hilt. The *Säbel* became the most common category of Nazi sword, with over 200 different styles being made available for private purchase by army officers alone.

Most sabres and rapiers were adorned with fancy and purely decorative coloured knots. These traced their ancestry back over 200 years to the simple leather straps which had been wrapped around the hilt and wrist to prevent such weapons being dropped and lost in combat. The knot for officers and senior NCOs was usually referred to as a Portepée, or 'sword carrier', while the plainer and normally stemless junior NCOs' and enlisted men's version was generally called a *Faustriemen*, or 'fist strap'.

Not only did Nazi swords vary greatly in design, they also varied considerably in length since tall men needed longer weapons than their shorter colleagues. The following table indicates the recommended *Schwert*, *Degen* and *Säbel* lengths for wearers of different heights. All sizes shown are in centimetres.

| HEIGHT OF WEARER | LENGTH OF *SCHWERT* | LENGTH OF *DEGEN* | LENGTH OF *SÄBEL* (OFFICERS) | LENGTH OF *SÄBEL* (NCOs & ENLISTED MEN) |
|---|---|---|---|---|
| 160 | 88 | 90 | 86 | 90 |
| 162 | 88 | 90 | 88 | 92 |
| 165 | 88 | 90 | 90 | 94 |
| 168 | 88 | 90 | 92 | 96 |
| 170 | 93 | 95 | 94 | 98 |
| 173 | 93 | 95 | 96 | 100 |
| 175 | 93 | 95 | 98 | 102 |
| 178 | 93 | 100 | 100 | 104 |
| 180 | 98 | 100 | 102 | 106 |
| 185 | 98 | 100 | 104 | 108 |
| 190 | 98 | 100 | 106 | 110 |
| 195 | 98 | 100 | 108 | 112 |

It is interesting to note that while swords and rapiers came in only three standard sizes, the ubiquitous sabre was produced in no fewer than fourteen assorted lengths for officers, NCOs and enlisted men. It was therefore possible to buy any given individual style of army sabre in that number of sizes, resulting in at least 2,500 military *Säbel*

variants being produced simultaneously for private purchase in the years before 1942, when manufacture began to decline. By that time, well over 5 million Nazi sabres were in circulation.

This chapter describes a representative selection of these weapons, together with the swords and rapiers worn by other Third Reich organisations.

# ARMY

Throughout the 1933–45 period, the German Army favoured the use of sabres. Inherited swords, passed

**103**. *Styles of hanging knot.*
**Top row** *(left to right): 1 – Bayonet* Troddel *for enlisted men; 2 – Sword* Faustriemen *for enlisted men; 3 – Sword* Faustriemen *for NCOs; 4 – Bayonet* Troddel *for NCOs*
**Labelled parts for top row:** *A – Cloth strap; B – Slide; C – Stem; D – Crown; E – Ball; F – Leather strap*
Bottom row (left to right)*: 5 – Dagger portepée for officers; 6 – Sword portepée for officers*
**Labelled parts for bottom row:** *A – Cord; B – Slide; C – Stem; D – Crown; E – Ball; F – Leather strap*

**104**. *Regulation methods of attaching hanging knots. 1 – Sword* Faustriemen; *2 – Bayonet* Troddel; *3 – Sword portepée; 4 – Dagger portepée (Luftwaffe style for short 23 cm knot)*

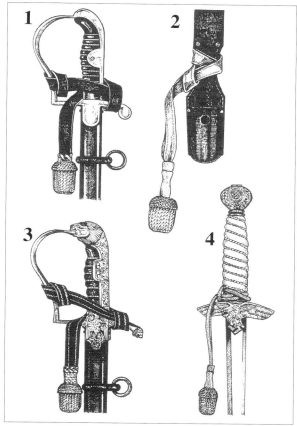

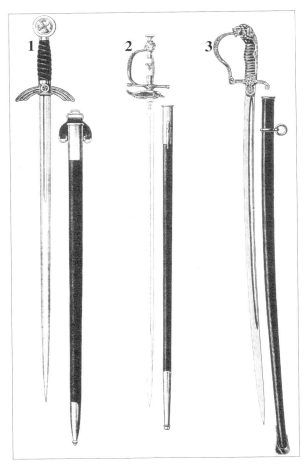

**105**. *Sword categories: 1 – Schwert,* or sword *(*Luftwaffe *1934 sword as example); 2 – Degen,* or *rapier (*Luftwaffe *general's 1935 degen as example); 3 – Säbel,* or sabre *(Army lion-head sabre as example)*

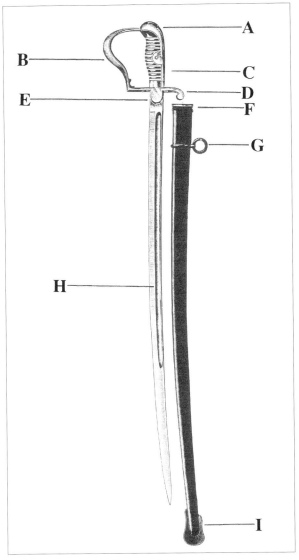

**106**. *Sword nomenclature (Army ordnance sabre as example): A – Pommel; B – Knuckle bow; C – Hilt; D – Crossguard/quillon; E – Langet; F – Throat; G – Carrying ring; H – Fuller/blood groove; I – Drag*

from father to son, were only very rarely permitted to be worn instead of the regulation *Säbel*, and the latter predominated on almost every occasion. Although sabres were produced in an enormous variety, all fell into one of two readily distinguishable groupings: ordnance and private purchase.

## THE ORDNANCE SABRE

The standard ordnance sabre, or *Einheitssäbel*, was issued on a unit basis for retention and return as required, and remained Army property. It was very simple in design, typically featuring a black plastic grip wrapped with wire, unadorned nickel fittings, a plain steel blade and a black stove-enamelled

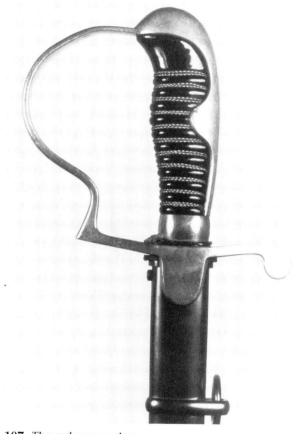

**107**. *The ordnance sabre.*

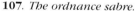

**108**. *An Army enlisted man posing with his ordnance sabre, 1938.*

scabbard. Each example was usually stamped with a *Waffenamt*, or armoury, inspection mark, and an accountability serial number. The *Einheitssäbel* could be carried by all officers and senior NCOs, irrespective of their branch of service, and by junior NCOs and enlisted men attached to cavalry regiments.

The junior NCO's and enlisted man's sabre was known as the *Mannschaftssäbel* and was a heavy-duty weapon weighing around 1.8kg, including scabbard. It was constructed using nickel-plated steel fittings, had a flat-backed blade and was worn with a simple plain-strap *Faustriemen*. The officer's and senior NCO's version of the ordnance sabre,

the so-called *Offizier-Einheitssäbel*, was much lighter in total weight at approximately 1.1kg. It featured silvered brass rather than steel hilt components, had a round-backed blade with a proportionately longer tip, and was accompanied by a more substantial portepée knot bearing two or three rows of metallic threads woven into the leather strap.

Regulations prescribed that the *Säbel* was to be worn on formal occasions with service, guard, reporting and undress uniforms. Officers and senior NCOs normally suspended their sabres from a leather hanger attached to a belt hidden under the tunic, while other ranks carried them from the external waist belt. Frogs were also supplied for hanging the *Säbel* on the left or right side of the saddle, by mounted officers and other ranks respectively. Ordnance sabres were regularly inspected by unit armourers and had to be cleaned after use, with dirt being removed, rust spots oiled and scabbards dried out. Only then could they be returned to storage in the lockers of the personnel to whom they had been issued.

Manufacture and distribution of the *Einheitssäbel* was discontinued on 20 December 1940 for the duration of the war.

## PRIVATE-PURCHASE SABRES

Many German Army officers and senior NCOs had chosen to wear a variety of privately purchased or *extra* sabres since the imperial era, and on 23 April

1934 Hitler officially sanctioned this practice. An extensive selection of non-regulation weapons subsequently developed, ranging from medium-quality silver-plated examples of the simple *Einheitssäbel* to highly ornate and expensive fire-gilded pieces. The style of *Extra-Säbel* most commonly favoured featured golden fittings, a lion-head pommel, an eagle and swastika crossguard and a nickel-plated blade, and was worn with a standard hanger and knot. However, virtually every sword-making firm offered its own distinctive patterns of sabre and so the variations on this theme were almost endless. Some examples had leopard, panther, puma or eagle-head pommels, with or without coloured glass eyes, while others bore crossed swords,

*110. This lion-head sabre, by Emil Voos, features the NSDAP-pattern national emblem and is hung with a regulation leather-strapped portepée knot.*

*109. A typical Army lion-head sabre, by the Puma company, with the* Wehrmacht-*style eagle and swastika on the langet.*

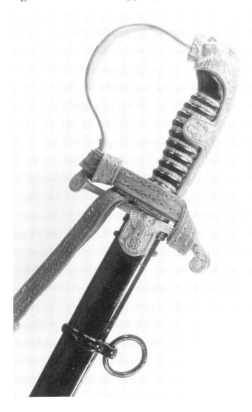

**111**. Generalfeldmarschall *Werner von Blomberg (right) in conversation with Italian Marshal Pietro Badoglio during Mussolini's state visit to Germany, September 1937. Von Blomberg wears the Alcoso-made Sword of Honour presented to him by the Army High Command on 13 March, 1937 to mark his fortieth year of military service.*

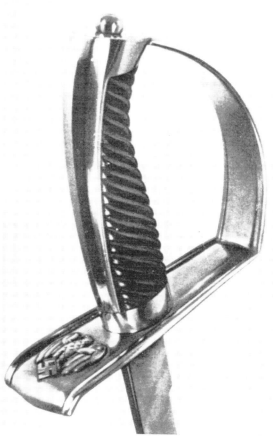

**112**. *Design drawing for an Army sabre with wooden grips and stainless steel fittings, published in 1941 but never sanctioned for manufacture.*

cannons, flags, swastikas, oakleaves or laurel branches on their crossguards. Such design details could be cast, die-struck or engraved, and fittings might be in steel, brass, bronze or aluminium, and be plated, painted or lacquered. Plain, etched, damascene or blued and gilded blades were also options. Many companies patented their own particular sabre patterns, which were then marked with the protective *Ges. gesch.* or '*DRP*' designations. Carl Eickhorn, for instance, produced the popular Field Marshal series of private-purchase

sabres with distinctive designs named after old German military heroes, including Blücher, Derfflinger, Lützow, Prinz Eugen, Roon, Scharnhorst, Freiherr von Stein, Wrangel and Zieten. Regardless of style, the reverse langet of each *Extra-Säbel* usually took the form of a shield-shaped or oval cartouche upon which was engraved the owner's initials.

The prices of private-purchase sabres varied considerably, depending upon the quality of workmanship involved. During 1939, when an average Army

*Leutnant* was receiving a salary equating to 60 Reichsmarks per week, the following typical prices were quoted in sales catalogues:

|  | Reichsmarks |
|---|---|
| Silvered sabre with plain fittings | 10.50 |
| Gilded sabre with plain fittings | 14.65 |
| Gilded sabre with engraved fittings | 16.50 |
| Silver-plated lion-head sabre | 16.50 |
| Gilded lion-head sabre | 17.25 |
| Gold-plated lion-head sabre | 22.25 |
| Gold-plated lion-head sabre with coloured glass eyes | 22.80 |
| Gold-plated lion head sabre with etched blade | 23.60 |
| Fire-gilded lion-head sabre with etched blade | 38.60 |
| Fire-gilded lion-head sabre with Damascus steel blade | 90.00 |

Manufacturers also offered a service repairing damaged sabres, and the following few examples give an idea of the charges involved:

|  | Reichsmarks |
|---|---|
| To remove a dent in a scabbard | 0.25 |
| To nickel-plate a blade | 1.00 |
| To re-enamel a scabbard | 2.10 |
| To shorten a sabre | 3.60 |
| To polish a sabre | 6.05 |
| To renew gold plating on a sabre | 8.65 |

A Field Marshal's sword and a number of innovative Army *Säbel* designs featuring stainless steel fittings, basket-style hilt guards and wooden grips were proposed in the Solingen Chamber of Commerce publication *Blanke Waffen* ('Edged Weapons') in 1941, but were never approved.

The production of private-purchase sabres finally halted on 27 May 1943, for reasons of economy.

## NAVY

The Navy sabre, or *Marine-Offiziersäbel*, although very richly decorated, was much more standardised than its Army equivalent and was one of the few edged weapons of the Third Reich not to bear the swastika as part of its exterior design. With the exception of the omission of the imperial crown

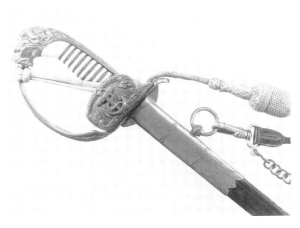

**113**. *The Navy sabre.*

after 1918, the sabre style remained basically unchanged from that of the *Kaiserliche Marine* 1848-pattern *Säbel* and commonly featured a gold-plated brass hilt with a white celluloid grip bound in gilt wire. The lion-head pommel had eyes of red and green glass, representing a ship's port and starboard warning lights, while the ornamentation of the spring-loaded basket-style handguards varied only slightly according to manufacturer. The large oval front guard was generally decorated with the traditional and international fouled anchor insignia, the smaller rear guard being plain and bored with a hole which fitted over a button to secure the weapon to its scabbard. The latter was constructed from black leather with finely engraved golden fittings and had two carrying rings attached to the upper and centre mounts. A variety of nickel-plated steel blades etched with different nautical motifs including battleships, destroyers, submarines, anchors, dolphins and so on was available, and small unobtrusive swastikas often appeared as integral parts of this decoration after 20 April 1938. Damascus steel blades were also produced for those who could afford to buy them. The regulation sabre was priced at around 31 Reichsmarks, while one with an ivory grip typically cost 42 Reichsmarks and an example with a Damascus blade 110.

The *Marinesäbel* was authorised for dress use by all ranks from *Feldwebel* upwards, with the blue uniform. It was never carried during the performance of routine daily duties on board ship or within barracks, except by special order. The weapon was suspended from two leather straps hung from an internal belt, and was accompanied by a 50 cm aluminium or gilt wire portepée knot looped around the upper part of the knuckle bow and then laced about the grip in a sailor's knot known as a *Webeleinensteg*. When the field-grey naval outfit was worn, by coastal artillery staff and other shore-based marine personnel, the sabre could be held within a more appropriate black stove-enamelled Army-pattern scabbard.

*114. This Navy issue sabre clearly exhibits the pre-May 1934 Kriegsmarine Waffenamt inspection stamp alongside accountability markings denoting weapon no. 1363 as listed on the 'Nordsee' area armoury inventory.*

Officers were expected to equip themselves with sabres at their own expense, while senior NCOs could draw them as required from unit stores. Issue examples were stamped with the usual accountability numbers alongside the *Kriegsmarine* property and inspection mark of an eagle and swastika over the letter 'M'. Those who had served in the First World War were allowed to continue sporting their imperial sabres throughout the Third Reich. However, inherited weapons could be carried only with the express permission of the Commander-in-Chief of the Navy, which was not normally granted unless the sabre concerned had actually been used by the wearer's direct ancestors in the face of the enemy.

Production of the *Marine-Offiziersäbel* ceased on 27 May 1943.

# AIR FORCE

Hermann Göring was fascinated by edged weapons, and by swords in particular, and was rarely seen in public without one at his side. There was nothing he liked better than to be presented with a unique bladed sidearm, a romantic weakness which was exploited to a great extent by organisations and individuals who wished to curry favour with Hitler's powerful heir-apparent. The *Luftwaffe* officer corps initiated the practice by gifting an ornate custom-made *Schwert* to Göring on the occasion of his wedding to the actress Emmy Sonnemann in April 1935, and the leaders of the Hunting Association, the Forestry Service and German industry, all of which Göring also headed, soon followed suit. By 1943, the *Reichsmarschall's* numerous palatial residences at Carinhall, Berchtesgaden, Berlin and elsewhere were bedecked with vast arrays of daggers and swords of all types, from priceless medieval Japanese *katanas* and jewelled Arab scimitars to Soviet Cossack *shasquas* and other common trophies of war, which had been showered upon him by sycophantic foreign dignitaries, military commanders and members of the Nazi political hierarchy. It is perhaps not surprising, therefore, that from the outset Göring took pains to influence the design of the swords and rapiers carried by the staff of his beloved *Luftwaffe* personally.

## THE *LUFTWAFFE* 1934 SWORD

In March 1934, at the behest of Göring, *DLV-Präsident* Bruno Loerzer introduced a very impressive sword for wear by officers and senior NCOs of the quasi-military *Deutscher Luftsport Verband* with parade, walking-out and service dress uniforms. This weapon, which was duly adopted by the *Luftwaffe* on 26 February 1935, was the only *Schwert*-pattern sidearm to see widespread use during the Third Reich. It was very similar in design to the *DLV* 1934 and *Luftwaffe* 1935 daggers, with both its wooden grip and aluminium scabbard being covered in high-grade dark blue Morocco leather. The steel blade was heavily nickel-plated,

and the circular pommel and down-swept wing crossguard featured inlaid brass sunwheel swastikas. All metal fittings were initially in nickel silver, then polished aluminium from 1936. The *Luftwaffe* sword was worn without a portepée knot, and was suspended from a permanently affixed blue leather hanger attached to a belt slung under the tunic. Its principal manufacturers were Eickhorn, Horster, Klaas, Malsch, Pack, SMF, Weyersberg and WKC.

The *Fliegerschwert*'s high-quality construction resulted in it being the most expensive of all general-issue Nazi swords, at a price of around 32 Reichsmarks. While many officers purchased

**115**. *Hermann Göring with a captured Cossack* shasqua *presented to him by the Commander-in-Chief of the Army,* Generalfeldmarschall *von Brauchitsch, during the autumn of 1941.*

**116.** Luftwaffe *1934 sword, complete with carrying harness and storage bag. The harness supported the weight of the weapon and was worn across the right shoulder, under the tunic, with the leather hanger protruding through a slit cut below the flap of the tunic skirt's left pocket.*

**117.** *The blade of this* Luftwaffe-*issue sword has the* Air Force *Waffenamt* inspection mark stamped alongside the finely etched maker's logo of Paul Weyersberg.

their own, sufficient quantities were kept in local unit stores for *ad hoc* issue to those who chose not to do so. Pieces retained as *Luftwaffe* property had their blades stamped with the Air Force armourer's *Waffenamt* inspection mark. They also bore an appropriate unit designation engraved under the crossguard and occasionally on the scabbard locket. Examples of such abbreviations included 'I/Flak 6' (1st Battalion, 6th Anti-Aircraft Regiment) and 'Fl.Gr.(ZA) Rotenburg' (Training Detachment, Rotenburg Air Group). A small number of swords were presented to popular officers by their colleagues or staff as celebratory or retirement gifts, and such items often had the blades etched with appropriate commemorative inscriptions.

Production of the *Fliegerschwert* ended on 25 March 1943, and its further use was prohibited on 23 December 1944.

It is interesting to note that the *Luftwaffe* 1934 sword was colloquially known by contemporaries as the *Deutscher Ordensschwert*, or 'Teutonic knight's sword', because of its distinctively medieval appearance.

### THE *LUFTWAFFE* GENERAL'S 1935 SWORD

When he left the Army Flying Corps at the end of the First World War, albeit as a national hero and holder of the coveted Order Pour le Mérite, Hermann Göring held the comparatively lowly

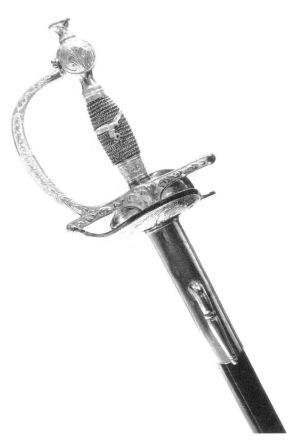

**118**. *The* Luftwaffe *general's 1935 sword was fashioned after the* Galadegen *worn at the old Prussian royal court.*

**119**. *Blade-etching on the* Luftwaffe *general's 1935 Sword of Honour.*

rank of *Hauptmann*. In August 1933, however, President von Hindenburg 'reactivated' Göring's military career in a most spectacular way, bestowing upon him the rank of *General der Infanterie*. By that time Göring had also been appointed by Hitler to the post of Prussian Minister of the Interior, which entitled him to wear the traditional Prussian Sword of Honour, a finely embellished *Stichdegen*, or fencing rapier, issued in very limited numbers and carried on ceremonial occasions by senior Prussian military and civil service officials. Göring liked the *Stichdegen* design so much that in March 1935 he decided to use it as the basis for a new and exclusive sword to be worn by his own Air Force generals.

Produced by Eickhorn and WKC, the standard *Flieger-Generalsdegen* had to be privately purchased by all *Luftwaffe* officers of the rank of *Generalmajor* and above. It was distinguished by its elegant hilt combination of a round pommel, slim knuckle bow, thin crossguard and oval shell guard, each of which was fire-gilded and decorated with acanthus leaves. The wooden grip was bound entirely in silver wire and bore a gold *Luftwaffe* eagle insignia. The plain-bladed weapon was accompanied by a silver bullion portepée knot, and was carried in a black leather scabbard with gilt brass fittings, suspended from a frog fixed to a belt worn under the tunic.

A special version of the *Generalsdegen* was presented by Göring, at his own expense, to selected *Luftwaffe* generals as a Sword of Honour. It equated to the Honour Daggers of the *SA*, *SS* and other Nazi political organisations, and had the blade handsomely etched in blue and gold with the

**120**. *A Luftwaffe general's 1937 sword. It is interesting to note that the Air Force eagle on the shell guard flies to the viewer's left, whereas that on the grip of the 1935 version flew to the right, the usual heraldic style.*

**121**. *Göring presenting the Spanish Cross to Lufthansa crews during the summer of 1939. He wears the* Luftwaffe *general's 1937 sword.*

dedication *In dankbarer Anerkennung – Der Reichsluftfahrtminister, Hermann Göring* ('In grateful recognition – from the Air Minister, Hermann Göring'). Recipients were normally First World War veterans who had demonstrated particular merit in the building up of the *DLV* and the *Luftwaffe*.

### THE *LUFTWAFFE* GENERAL'S 1937 SWORD

In October 1937, an updated *Generalsdegen* was introduced which conformed more to the style of the new *Luftwaffe Offizierdolch* and also reflected Göring's increasingly powerful status as head of the rapidly expanding Air Force. Also made in standard and Honour patterns by Eickhorn and WKC, the hilt configuration was essentially the same as that of its predecessor but had a white or yellow celluloid grip without the *Luftwaffe* eagle, the latter being incorporated in a larger size on the shell guard. The revised form of Honour blade etching read *In dankbarer Anerkennung – Der Oberbefehlshaber der Luftwaffe* ('In grateful recognition – from the Commander-in-Chief of the Air Force'), followed by a facsimile of Göring's signature.

From July 1940, the Sword of Honour was frequently referred to as the *Ehrendegen des Reichsmarschalls*. The flamboyant Göring continued to bestow it gratis upon his favourite generals until the end of the Second World War.

**122**. *A Stabswache* Göring sabre, with the ubiquitous SA insignia on the langet. This symbol was designed in 1929 as the result of a contest sponsored by the Oberste SA-Führung, *and combined runic and Gothic versions of the letters 'S' and 'A'.*

# SA

In common with most other Nazi political organisations, the *SA* in general was not authorised to use a regulation sword. Only two specific elements of it were allowed to do so, the *Stabswache* Göring and the *Feldjägerkorps*. Ostensibly, both units carried swords because of their closely related Police-type functions. However, it is perhaps more significant from the sword point of view that both also came under the direct control of Hermann Göring, who doubtless exerted his considerable influence to provide his personal troops with appropriate edged weapons.

### THE *STABSWACHE* GÖRING 1933 SABRE

In January 1931, *SA-Stabschef* Ernst Röhm set up a number of *Stabswachen,* or Staff Guards. These constituted a reliable bodyguard of armed officers and men of at least one year's service in the *SA* or *SS*, who were made responsible for the protection of particular individuals and those members of their staff considered to be of critical importance to the National Socialist movement. Hermann Göring was allocated a company-sized unit known as the *Stabswache* Göring.

During 1933, Göring issued the few officers in his formation with an Army-pattern sabre to which was attached a red portepée knot with grey edge stripes and a red *SA* insignia on its silver-grey stem. The only known photograph depicting this *SA-Säbel* and knot being carried dates from 7 December 1933. Due to the fact that the *SA* Staff Guards were armed, they came to be regarded as part of the anti-Hitler 'second revolution' threat posed by Röhm. After the latter's fall in the Night of the Long Knives, the *Stabswachen* were quickly disbanded and their most loyal personnel were incorporated into the *Feldherrnhalle Standarte*. Consequently, use of the *Stabswache* Göring sabre was discontinued at the beginning of July 1934.

### THE *FELDJÄGERKORPS* 1933 SABRE

In February 1933, Hermann Göring ordered the institution of an Auxiliary Police Force, or *Hilfspolizei*, raised from trusted members of the *SA* and *SS*, to support the regular *Schutzpolizei* in Prussia's larger cities. On 7 October 1933, the *Hilfspolizei* was reorganised into a new formation, the 1,600-strong *Feldjägerkorps Preussen* (*FJK*), with eight *Abteilungen* located at Berlin, Breslau, Düsseldorf, Frankfurt-am-Main, Hanover, Königsberg, Magdeburg and Stettin. *FJK* troops were classed as superior to all other *SA* and *SS* men, regardless of rank, and they were invaluable in ensuring that the politically unreliable elements within the *Schutzpolizei* were identified and weeded out at an early stage.

In order to reinforce their authority in symbolic terms, Göring issued *FJK* officers with an Army-pattern sabre, the langet of which bore the *Feldjägerkorps* emblem of a six-pointed star surmounted by the imperial Prussian eagle with a swastika on its chest. The *FJK-Säbel* was accompanied by the same distinctive *SA* knot as that worn with the *Stabswache* Göring sabre.

On 1 April 1935, the *Feldjägerkorps* was wholly incorporated into the Prussian *Schutzpolizei* and adopted regular Police sidearms. The wearing of the *FJK* sabre ceased thereafter.

# SS

In 1934, Heinrich Himmler laid down the principle that all *SS* men were entitled to demand satisfaction with pistol or sword for affronts to their honour and integrity. In reality, special *SS* courts of honour usually intervened to prevent internal disputes proceeding to actual duels, particularly as Hitler had prohibited this anachronistic practice in the armed forces. However, from 1935 the *SS* was the only Nazi political organisation whose leader saw fit to accord its members the right to wear swords as traditional symbols of honour and authority. This distinction acknowledged not only the naturally close relationship between Himmler's forces and the regular Police, but also the continually developing rapport between the *SS* and the *Wehrmacht*.

The earliest use of *SS* swords was restricted mainly, although not entirely, to ranking personnel of the armed *SS* formations, namely the *SS-Verfügungstruppe* (*SS-VT*), the *SS-Totenkopfverbände* (*SS-TV*), and the *SS-Junkerschulen,* or officer training schools. They provided valuable military experience for many officers and NCOs who were later to become prominent personalities in the divisions of the *Waffen-SS*. A range of different *SS* swords was produced between 1933 and 1945, and these are best described in turn.

## SS CAVALRY SABRES

A *Berittene SS-Abteilung*, or mounted *SS* detachment, was set up in Munich as early as January 1931, and by March 1935 there were twenty-three *Allgemeine-SS* cavalry regiments, or *Reiterstandarten*, spread across Germany. Several of their headquarter towns were former garrisons of imperial cavalry units and consequently had excellent equestrian facilities readily to hand. Moreover, nationalist riding clubs were incorporated 'lock, stock and barrel' into the *Allgemeine-SS* during the 1930s, bringing with them their expertise in horsemanship. All this meant that the *SS Reiterstandarten* soon became acknowledged as being amongst the best cavalry formations of the Third Reich.

During peacetime, the *SS* cavalry were always ceremonial in function, with a distinctly snobbish outlook, and were seldom if ever deployed to assist the Police in practical domestic crowd-control situations. However, after the outbreak of war in September 1939 the majority of *SS* horsemen suddenly found themselves conscripted into Army cavalry units for combat service, or into the *SS-Totenkopfreiterstandarten* which had been hastily mustered to perform security duties behind the front lines. The latter units duly combined to form the *SS-Kavallerie-Brigade*, and subsequently earned an unsavoury reputation for the brutal treatment of civilians and other 'soft targets' in the occupied eastern territories. By 1942 an entire *SS-Kavallerie-Division* had been activated for counter-guerrilla operations in Russia, and it was honoured with the name 'Florian Geyer' in 1944.

All of these equestrian *SS* formations were commanded during the various stages of their development by Hermann-Otto Fegelein, whose ever-strengthening position in Nazi circles culminated in his marriage on 3 June 1944 to Gretl Braun, sister of Hitler's mistress. The 38-year-old *SS-Gruppenführer* Fegelein attempted to flee Berlin in the last days of the war and was subsequently shot, on the *Führer*'s personal orders, for desertion.

*SS* cavalrymen of all ranks were issued with the Army-pattern ordnance sabre for wear, instead of the service dagger, when on horseback or walking out. Its plain black and silver design well matched not only the standard *Allgemeine-SS* uniform, but also the black outfit worn by members of the *SS-VT* and *SS-TV* on ceremonial occasions. During the 1933–5 period, a small number of unofficial *Einheitssäbel* variants were produced with Sigrunes etched, engraved or enamelled on the obverse langet, while a few cavalry officers and senior NCOs kitted themselves out with more elaborate sabres bearing silver lion-head pommels. These did not accord with *SS* regulations and were soon forbidden by Himmler, but they continued to be tolerated locally in the short term. Such early *SS* sabres were usually accompanied by the standard Army officer's portepée knot, or the Army *Faustriemen*, depending upon the wearer's rank. In

1935, however, with the increasing standardisation of *SS* uniform, these accoutrements were replaced by three distinctive *SS*-pattern sword and bayonet knots. The new sword portepée for all *SS* officers and warrant-officer-grade senior NCOs was made entirely of aluminium wire, with double black edge stripes to the strap and black *SS* runes woven into the stem. Its counterpart for junior NCOs was essentially very similar, but was distinguished by black flecking to the slide and black vertical lines to the acorn. The third knot, for enlisted ranks of the *SS* cavalry, had a black strap with double white edge stripes, a black slide, black runes on the white stem and a black crown to the white acorn. The latter knot was also worn on the bayonet by enlisted ranks of other *SS-VT* and *SS-TV* units until 1937.

*SS* cavalry sabres were withdrawn from general

**124.** Reichsführer-SS *Heinrich Himmler (left foreground) converses with* SS-Obergruppenführer *'Sepp' Dietrich, commander of Hitler's armed SS bodyguard, in September 1935. While Himmler and the others carry regulation* SS *1933 daggers, Dietrich's military status is emphasised by his silver-plated lion-head sabre.*

**123.** *An* SS *lion-head sabre by WKC, with Sig-runes cast into the obverse langet and the original owner's initials 'F.C.' engraved on the reverse. The pommel is embellished with red glass eyes, while the style of knot shown was commonly used by* SS *officers during 1934–5.*

use during 1939–40 for the duration of the war.

## SS OFFICERS' AND NCOS' SABRES, 1933–6

Between 1933 and June 1936, career officers and senior NCOs of the *SS-VT*, *SS-TV* and *SS-Junkerschulen* reinforced their military status by purchasing non-regulation Army-style sabres which they wore with service and undress uniforms on formal occasions. The only common denominator was usually a silver-plated lion-head pommel, with a selection of miscellaneous embellishments being encountered, including etched blades and the addition of Sig-runes to the reverse langet. These weapons were again worn with Army knots until the autumn of 1935, after which time the new *SS* portepées were adopted instead. Prior to the introduction of the standard *SS* officer's and NCO's swords in 1936, sabres readily distinguished ranking personnel of the elite armed *SS* from the mass of their colleagues in the non-equestrian units of the *Allgemeine-SS*, who continued to sport daggers as their designated sidearms.

## SS PRESENTATION SWORDS, 1933–6

Before the tightening of regulations governing the wearing of *SS* edged weapons in June 1936, specially commissioned swords were occasionally presented to well-respected *SS-VT* officers by their staffs as celebratory gifts. These rare items, designed to individual specification, generally had their blades inscribed with suitable commemorative dedications. In October 1935, for example, to mark his transfer from the *SS-VT* to an Army alpine regiment, former First World War air ace *SS-Obersturmbannführer* Georg Ritter von Hengl received a sword appropriately similar in style to the new *Luftwaffe* general's *Stichdegen*, with runes on the pommel. Its blade was etched with the legend: *Ihrem Kommandeur zur Erinnerung – Das Führer-Korps I/SS 'Deutschland'* ('In remembrance of our commander – the officer corps, 1st Battalion, *SS*-Regiment 'Deutschland'). Sepp Dietrich, the charismatic leader of Hitler's *SS* bodyguard regiment, was presented with an exceptionally elaborate eagle-head *Degen* featuring hallmarked silver fittings in May 1936. Its extremely expensive Damascus steel blade bore the raised silver inscription *Dem Kommandeur der Leibstandarte SS Adolf Hitler*, along with the names of no fewer

126. Waffen-SS *recruits take the oath of loyalty on the blade of an* SS-Ehrendegen, *1943*.

than 105 officers who had contributed towards its purchase. Although Himmler categorically forbade the further use of such unofficial *SS* presentation swords the following month, Dietrich defiantly wore this cherished weapon during ceremonial events at the Berlin Olympic Games that August, after which it was reluctantly consigned to storage in the basement of his home.

## THE *SS* OFFICER'S 1936 SWORD

On 21 June 1936, at the same time as he authorised the *SS* chained dagger, Himmler introduced a standard sword to replace the range of non-regulation lion-head sabres and unofficial presentation pieces hitherto worn by *SS* officers. Designed by Karl Diebitsch and manufactured principally by Peter Krebs of Solingen, the impressive new weapon was in the classic straight-bladed *Degen* style with a plain C-shaped knuckle bow. Its black wooden grip was wrapped in polished silver wire and was set with large silver *SS* runes within a circle. The raised pommel took the form of a sunburst, while the ornately cast ferrule comprised vertical oakleaves and acorns. Each hilt was stamped with the *SS* proofmark of two superimposed Sig-runes within an octagon, denoting that the overall design had been commissioned and approved by the *SS*. The sword's black stove-enamelled scabbard had Germanic ornamentation to its silver-plated locket, which also tended to be proof stamped, and an embossed silvered chape held in place by two

125. *An* SS *officer's 1936 sword by Peter Krebs, with regulation knot and silver bullion hanger. This so-called* Ehrendegen, *or Sword of Honour, was given an elevated status by Himmler, and was greatly prized by those who received it.*

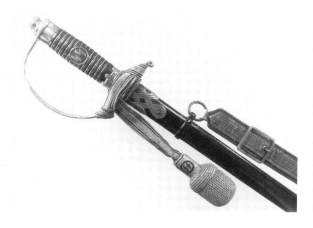

protruding screws. The *Degen* was worn with a black leather or silver bullion hanger, and was finished off with the regulation *SS* aluminium portepée knot.

From the outset, the new officer's sword was given a much elevated status and was referred to as the *Ehrendegen des Reichsführers-SS*, or 'Reichsführer's Sword of Honour'. It could not be privately purchased or worn automatically by every *SS* officer, but was conferred by Himmler only upon:

- specially chosen full-time or honorary *Allgemeine-SS* leaders, particularly those on his personal staff or with military backgrounds, in recognition of long or meritorious service
- selected *SS-VT* and *SS-TV* senior officers, in acknowledgement of military achievement
- all new graduates of the elite *SS-VT Junkerschulen* at Bad Tölz and Braunschweig

Each presentation of the *Ehrendegen* was accompanied by a signed citation in which the *Reichsführer* instructed the recipient: *'Ich verleihe Ihnen den Degen der SS. Ziehen Sie ihn niemals ohne Not! Stecken Sie ihn niemals ein ohne Ehre!'* ('I award you the *SS* sword. Never draw it without reason, or sheathe it without honour!') This exhortation was copied from the Castilian blade motto *'No me saques sin rason – no me embaines sin honor'*, popular amongst Toledo sword and dagger manufacturers in the 17th century. Bestowals of the *Ehrendegen* were recorded in the *SS Dienstaltersliste*, or Officers' Seniority List, which reveals that only 86 per cent of even the highest-ranking *SS* commanders (colonels and above) were entitled to wear it. That percentage can be broken down as follows:

| | |
|---|---|
| *Standartenführer* | 58% |
| *Oberführer* | 83% |
| *Brigadeführer* | 90% |
| *Gruppenführer* | 91% |
| *Obergruppenführer* | 99% |
| *Oberst-Gruppenführer* | 100% |

Consequently, the sword was very highly prized by those junior and middle-ranking officers fortunate enough to receive it.

*SS* cavalry officers who were presented with the *Degen* could wear it instead of the regulation *SS-Säbel*, while recipients holding dual *SS* and Police commissions were permitted to sport the *Reichsführer*'s Sword of Honour with the Police as well as the *SS* uniform. During the war, *Waffen-SS* recruits were frequently sworn in at ceremonies where the oath of loyalty was taken while placing hands on the unit commanding officer's sword blade, a custom harking back to the Middle Ages. Any officers dismissed from the *SS* were obliged to return their *Degen* to Himmler. However, the swords of those who were honourably discharged, or who retired or died in service, could be retained and passed on to their next-of-kin.

Manufacture of the *Ehrendegen* was discontinued on 25 January 1941 for the duration of the war. *Waffen-SS* officers commissioned after that date frequently reverted to the old practice of carrying miscellaneous Army lion-head sabres decorated with *SS* knots.

**127.** *An SS NCO's 1936 sword, carried by warrant officers of the* SS-VT.

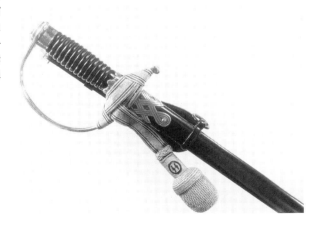

## THE SS NCO'S 1936 SWORD

A sword for wear with parade and walking-out dress by senior *SS-VT* NCOs, ranking between the warrant-officer grades of *Oberscharführer* and *Sturmscharführer*, was also created on 21 June, 1936. It was very similar in design to the officer's *Ehrendegen*, but had heavier nickel-plated steel fittings, a simple black wooden grip without wire wrap or insignia, a flat pommel bearing the *SS* runes, and a plain black lower scabbard with an integral drag rather than a silvered chape. The hilt and scabbard locket were usually *SS* proof-stamped, and the sword was accompanied by a regulation aluminium portepée knot.

Limited numbers of the *SS-Unterführerdegen* were retained at local *SS-VT* unit stores for issue and return as required. They could not be bought privately, and their wear was never extended to the *Allgemeine-SS*. Production ceased on 25 January, 1941.

## THE SS DAMASCUS SCHOOL AT DACHAU

One of the least known of the various economic enterprises run by the *SS* was the manufacture of exquisite Damascus steel blades. The art of Damascus forging was brought to Europe from the Middle East during the Crusades, and involved the continual folding of around 500 white-hot layers of steel and iron which were then immersed in a warm oil. The result was a blade of extreme strength and beauty, with a patterned surface. Damascus swords were popular amongst officers of most Western nations during the eighteenth and nineteenth centuries, even though they could cost up to thirty times as much as their standard-issue counterparts. By the early twentieth century, however, far cheaper versions with acid-etched blades giving the appearance of Damascus steel began to be produced, and the Damascus industry was further hit when many of its highly skilled craftsmen perished during the First World War. At the end of 1936 there were only six qualified Damascus swordsmiths in the whole of Germany: Robert Deus, Paul Dinger, Paul Hillmann, Otto Kössler, Paul Müller and Karl Westler. The majority of these individuals were quite elderly, and there was a real danger of their art dying with them.

At the beginning of 1937, Himmler engaged 54-year-old Paul Müller, the youngest of the surviving Damascus smiths, to forge an ornate patterned blade bearing the raised golden inscription 'In good times and bad, we will always remain steadfast'. It was incorporated into an upgraded version of the new *SS-Ehrendegen*, with hallmarked silver fittings specially produced by the *SS* jeweller Karolina Gahr. The resultant unique sword, extolling the virtues and loyalty of the *SS* officer corps, was presented to Hitler on 20 April that year, to mark the occasion of his forty-eighth birthday, and was extremely well received.

Himmler was thereafter anxious to preserve the traditions of German Damascus production, and in December 1938 he ordered the setting up of a Damascus school within the confines of Dachau Concentration Camp. The theory was that selected members of the *Allgemeine-SS* who were cutlers or blacksmiths by profession would be temporarily accommodated at Dachau and trained in Damascus forging at *SS* expense. The *Reichsführer* chose Müller as director of the enterprise, and a contract was drawn up between them.

Its terms were as follows:

- Müller pledged himself to teach designated *SS* men how to make Damascus blades, so that the art would be preserved and carried on by the *SS*.
- He had to comply with all directions of the *Reichsführer-SS*, and was responsible to him personally for his actions. He could be recalled from leave should Himmler wish a special presentation blade to be made urgently.
- He received an initial net monthly salary of 450 Reichsmarks, plus free accommodation.
- If he became unable to work or was restricted in his work because of an industrial injury, he would continue to receive his entire salary.
- On reaching the age of sixty-five, he would receive a life pension of 75 per cent of his salary at retirement.
- The contract was valid for Müller's lifetime.

Müller had recently fallen on hard times and he jumped at the chance of regular employment which would give him security in his old age. The contract was duly signed, and the *SS* Damascus School at Dachau went into production.

In 1939, the first ten students were assigned to the school and began to forge Damascus blades on Himmler's orders, for fitting to presentation swords destined for senior Nazi personalities. They also designed one of the three prototype *Waffen-SS* daggers produced in 1940. As the war dragged on, however, battlefield exigencies inevitably claimed Müller's assistants. By 1943 he had only two apprentices left, Doll and Noth, and one trained Damascus craftsman called Flittner. That April all three were drafted into the *Waffen-SS*, leaving Müller to operate the smithy alone. He objected bitterly to their enforced departure, but the response from *SS* Headquarters was unsympathetic and the situation had to be accepted as a *fait accompli*. Paul Müller remained at the Dachau smithy and continued to produce presentation blades for the *Reichsführer-SS* until the beginning of 1945.

### SS Presentation Swords, 1939–45

The most exclusive of all *SS* edged weapons were the so-called *Geburtstagsdegen*s, or 'birthday swords', given by Himmler to *SS* generals and other leading Nazis as birthday presents. Like the Adolf Hitler *SS* Sword of 1937, these featured hallmarked silver fittings by Gahr which were married to suitably inscribed Damascus steel blades from the *SS* smithy at Dachau. The *Degen* gifted to Foreign Minister and *SS-Obergruppenführer* Joachim von Ribbentrop on his birthday in 1939, for example, bore the golden legend *Meinem lieben Joachim von Ribbentrop zum 30.4.39 – H. Himmler, Reichsführer-SS* set between two swastikas. Other ornate *SS* swords were produced in ever-declining numbers until the end of the war, as a means of acknowledging outstanding merit. A typical recipient was *SS-Gruppenführer* Heinrich Müller, Chief of the Gestapo, who was given one of the last Damascus-bladed presentation swords early in

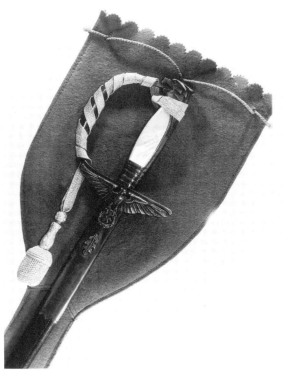

**128**. *A Diplomatic Corps sword by Alcoso, complete with its portepée knot and brown felt tie-neck storage bag.*

1945 in recognition of his work as head of the 20 July Investigation Commission which tracked down those responsible for the celebrated assassination attempt on Hitler's life the previous year.

Just as Göring encouraged the distribution of edged weapons to members of the organisations under his jurisdiction, there can be little doubt that the plethora of swords worn by the *SS* derived from Himmler's vision of his men forming a knightly order governed by Germanic traditions.

## DIPLOMATIC CORPS

In May 1938, at the same time as the introduction of the *Diplomaten-Dolch*, a court sword, or *Galadegen*, was authorised for wear with formal evening dress by senior members of the

Diplomatic Corps. Its silvered hilt featured a C-shaped knuckle bow but was otherwise identical to that of the diplomat's dagger, with a stylised eagle-head pommel, mother-of-pearl grips and a large left-facing eagle on the crossguard. The thin, straight blade had an ornate acanthus-leaf pattern etched on both sides, with an eagle and swastika positioned centrally on the obverse. A flat-strapped silver bullion portepée knot was wrapped around the handle. The weapon was held in a black leather scabbard with silver-plated fittings, the locket having a protruding lug in the shape of an oakleaf which facilitated suspension from a black leather frog attached to a shoulder harness.

The elegant *Diplomaten-Degen* was manufactured only by Alexander Coppel and Carl Eickhorn. Production ceased in July 1942.

# POLICE

During 1933, officers and warrant officers of the various localised police forces spread across Germany continued the well-established imperial and Weimar practice of wearing Army-pattern sabres, often bearing their own provincial coats-of-arms. The latter were generally superseded by the Nazi Police eagle and swastika insignia following the formation of the national *Deutsche Polizei* at the beginning of 1934, but use of the traditional *Säbel* as a symbol of Police authority was maintained for another two years. Indeed, Police cavalry units, like those of the *SS*, continued to be issued with the *Einheitssäbel* until 1940.

### THE POLICE OFFICER'S 1936 SWORD

On 21 June 1936, only four days after he had been nominated *Chef der Deutschen Polizei*, Himmler ordered the introduction of a range of Police swords to mirror the ever-closer relationship between the *SS* and the forces of law and order. The new Police officer's sword, or *Polizei-Führerdegen*, was identical to the *Ehrendegen des Reichsführers-SS* but the grip bore matt aluminium wire and a Police eagle rather than polished silver wire and the *SS* runes. Police officers who held *SS* membership had the option of

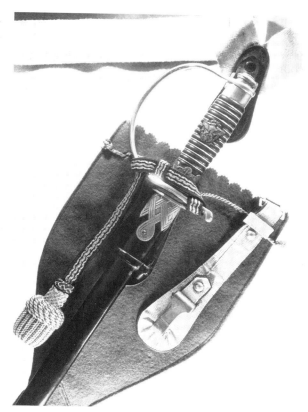

129. *A Police officer's 1936 sword by Hermann Rath, complete with all its hanging and storage accoutrements.*

sporting the *SS* runes engraved on the Police sword pommel. Those who simultaneously held *SS* commissions or warrants were expected to use the regulation *SS* knot instead of the standard Police portepée, which comprised an aluminium acorn with a black leather strap bearing red and silver thread decoration.

The *Polizei-Führerdegen* could be bought privately for around 19 Reichsmarks (20 if pommel runes were ordered) and was worn by all officers, *Meisters* and *Oberjunkers* of the Order Police with parade and walking-out uniform. Manufacture was discontinued shortly after the outbreak of the Second World War.

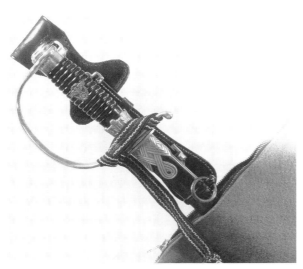

**130**. *A Police NCO's 1936 sword, with hanger, knot and grey felt zip-neck storage bag.*

## THE POLICE NCO'S 1936 SWORD

At the same time as the creation of the officer's sword, a *Polizei-Unterführerdegen* was authorised for wear by senior NCOs graded as warrant officers, i.e. *Hauptwachtmeister*, *Oberwachtmeister* and *Zugmeister* with at least nine years' service. The sword was very similar to the *SS-Unterführerdegen*, but featured a plain pommel and the Police eagle insignia set into the grip. Police NCOs who held *SS* membership could have the *SS* runes engraved on the pommel of the Police sword as an optional extra, while those simultaneously holding *SS* commissioned or warrant-officer rank were obliged to use the *SS* knot instead of the regulation Police portepée.

The Police NCO's sword was available for private purchase at 17.60 Reichsmarks (18.60 with pommel runes). Production ceased during 1939–40.

## BADGELESS SS/POLICE SWORDS

Two variants of the *SS* and Police officer's and NCO's *Degen* exist which do not have any form of insignia on their grips or pommels. These are often attributed to *SS* officer cadets and aspirant NCOs, but there is no reference to them in surviving *SS* uniform regulations and they may, in fact, have been intended for members of a Police-related organisation. To date, their true purpose remains unconfirmed.

The following table summarises the various permutations of standard *SS*/Police 1936-pattern swords authorised, and those entitled to carry them:

| SWORD/ACCOUTREMENT | WEARER |
|---|---|
| SS officer's sword with SS knot | Selected *Allgemeine-SS*, *SS-VT* and *SS-TV* commissioned officers; the sword could be worn with SS or Police uniform |
| SS NCO's sword with SS knot | All *SS-VT* and *SS-TV* warrant officers |
| Police officer's sword with Police knot | All Police commissioned officers and officer cadets |
| Police officer's sword with SS knot | Police commissioned officers and officer cadets simultaneously holding SS commissioned-officer or warrant-officer rank |
| Police officer's sword with SS runes to the pommel (optional) | Police commissioned officers and officer cadets holding SS membership |
| Police NCO's sword with Police knot | All Police warrant officers |
| Police NCO's sword with SS knot | Police warrant officers simultaneously holding SS commissioned-officer or warrant-officer rank |
| Police NCO's sword with SS runes to the pommel (optional) | Police warrant officers holding SS membership |
| SS/Police officer's sword without insignia to the grip or pommel | Unconfirmed |
| SS/Police NCO's sword without insignia to the grip or pommel | Unconfirmed |

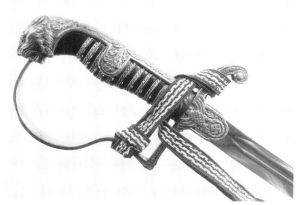

**131**. *A traditional Fire Brigade sabre of the type worn between 1933 and 1936, with a steel helmet, crossed axes and flaming torch emblem on the langet. The blade is etched with similar designs.*

**132**. *The Eickhorn model 1695 sabre, which was identical to the 'Wrangel' pattern but with a lion-head pommel, was worn by many Fire Brigade officers from May 1936.*

# FIRE BRIGADE

On 27 May 1936, local Fire Brigade officers of the rank of *Hauptbrandmeister* and above were required to discontinue further use of the 1870-pattern dress dagger and adopt a gilded plain-hilted *Einheitssäbel*, or a sabre of the Eickhorn Blücher or Wrangel model instead. The latter swords featured the eagle and swastika on the langet, replacing earlier fire-fighting emblems which were now expressly forbidden. At the same time, *Feuerwehr* NCOs and men had their dress axes superseded by unadorned bayonet-like *Faschinenmesser* measuring 35 cm and 40 cm respectively.

In 1938, all the regular provincial fire brigades throughout Germany were amalgamated to form the *Feuerschutzpolizei*, a new branch of Himmler's ever-expanding Order Police. From that time, professional Fire Brigade officers of the ranks of *Oberbrandmeister* and upwards were entitled to wear the standard 1936-pattern *Polizei-Führerdegen* with parade and walking-out uniform. Fire officers who were also members of the *SS* could use the runic pommel or *SS* knot, under the same conditions as their *Ordnungs-polizei* colleagues. Similarly, holders of the

*Ehrendegen des Reichsführers-SS* were expected to wear it in conjunction with the *Feuerschutzpolizei* outfit. NCOs were not authorised to carry the *Polizei-Unterführerdegen*, and they continued using the 35 cm nickel-plated *Faschinenmesser* on dress occasions.

Part-time voluntary personnel attached to the many auxiliary fire brigades, or *Freiwilligen Feuerwehr*, based in larger German towns were not classed as full members of the Police and were therefore not accorded the right to bear Police swords. Volunteer fire officers continued to sport gilded sabres after 1938, with bayonets being used by non-commissioned ranks.

# CUSTOMS SERVICE

Swords worn by officer-grade personnel of the Customs Service fell into two distinct categories, one for the *Landzoll* and the other for the *Wasserzoll*.

### LAND CUSTOMS SWORDS

Officials of the *Landzoll* were not given their own unique pattern of sword, but were allowed to use their discretion in choosing from a limited number of the many existing Army sabres on the market. The following styles were specifically approved by the Ministry of Finance and designated for wear with the Land Customs uniform:

- the Eickhorn model no. 40, which was a plain gold-hilted *Einheitssäbel*
- the Eickhorn Blücher model, which had a gold lion-head pommel and an open-winged eagle and swastika on the langet
- the WKC model no.1016, which was virtually identical to the Blücher sabre but with silver fittings
- the Eickhorn Freiherr von Stein model, which had a gold oakleaf pommel and a close-winged eagle and swastika on the langet

All of these sabres had plain black steel single-ring scabbards, and were worn with a green and silver portepée knot.

**133**. *This rare Water Customs sabre distributed by A. Lüneburg of Kiel is similar in style to the pattern patented by Eickhorn, but has a variant eagle to the langet. The lion-head pommel has red and green glass eyes. Note also the black leather hanging strap with lion-mask buckle.*

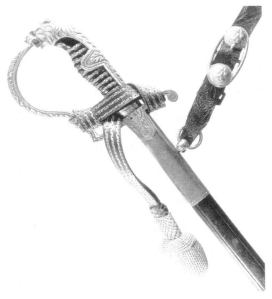

## WATER CUSTOMS SWORDS

*Wasserzoll* officials generally sported the naval officer's sabre and knot with their dress uniform. The Eickhorn firm alone patented a distinctive and hybrid *Säbel* which it marketed as being specifically for Water Customs use. This weapon took the form of an Army-pattern Blücher-model sabre married to a Navy-pattern black leather scabbard, with gilded fittings and two carrying rings. This Eickhorn *Säbel für Wasserzoll* had not been commissioned by the customs authorities, and was provided simply as a commercial alternative for marine-orientated buyers who preferred not to use the plain Army scabbard with what was, after all, a pseudo-naval uniform.

# CRIMINAL JUSTICE AND PRISON SERVICES

The *Justizdienst*, which administered criminal justice in the Third Reich, was strictly controlled by the Nazi party from the earliest days of Hitler's rule in Germany. Through the passing of legislation such as the Protection of People and State Act (21 February 1933), the National Labour Regulation Act (20 January 1934) and the Reich Reconstruction Act (30 January 1934), the *Führer* was able to centralise power, abolish the trade unions and legally gag his political opponents. On 24 April 1934, the sinister *Volksgerichtshof*, or People's Court, was set up as the highest tribunal in the land. It was presided over by two professional judges and five other officials carefully chosen from the ranks of the party, the *SS* and the armed forces, thus giving the latter group a majority vote. The court could impose the death sentence for a wide variety of crimes from treason and murder to 'serious breaches of the peace by armed persons', and there was no appeal against its decisions. In 1935, *NSDAP* membership was made obligatory for all senior personnel attached to the Ministry of Justice, including Supreme Court judges, subordinate magistrates and prosecuting lawyers. Even defence solicitors had to be approved by a Nazi vetting procedure before being permitted to represent accused persons during trials.

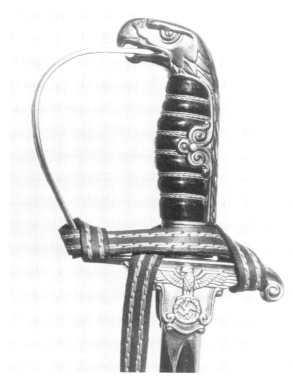

**134**. *A Justice Service 1936 sabre by Eickhorn, with distinctive eagle-head pommel.*

In addition to its prosecuting role, the Justice Ministry also supervised the extensive German prison system comprising the following types of establishment:

- *Zuchthaus*, for adults sentenced to penal servitude
- *Strafgefängnis*, for adults sentenced to imprisonment
- *Sicherungsanstalt*, for the detention of habitual or dangerous criminals after completion of sentence
- *Haftanstalt*, for the punishment of minor offenders
- *Jugendgefängnis*, for the punishment of youth criminals
- *Jugendarrestanstalt*, for the punishment of minor youth offenders
- *Arbeitshaus*, for the detention and education of

vagrants, prostitutes and so on after completion of sentence

So-called 'extraordinary prisons' existed for the specific purpose of detaining political and racial prisoners, but these fell outside the auspices of the *Justizdienst*, being run by Himmler's Security Police. They were operated under a regime of extreme brutality. The majority of their internees were German political offenders, military delinquents and Jews. Members of the *Reichsführer*'s forces were placed above the law and could be tried only in specially convened *SS* and Police courts for offences against their internal discipline code.

## THE JUSTICE SERVICE 1936 SABRE

In 1936, senior officials of the Criminal Justice Service were authorised to wear a unique style of Army-pattern sabre on ceremonial occasions. It featured an eagle-head pommel, an overlapping feather design to the knuckle bow and ferrule, and an eagle and swastika on both the obverse and reverse langets. The *Justiz-Säbel* had silver-plated fittings and was accompanied by an aluminium portepée knot.

**135.** *A Prison Service 1936 sabre. This variant example is the WKC model 1047, and features oakleaves to the ferrule rather than the usual overlapping feather design.*

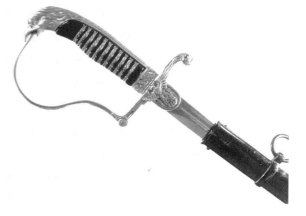

### THE PRISON SERVICE 1936 SABRE

A sabre identical to that allocated to the Justice Service, but with gold-plated hilt fittings, was approved in 1936 for wear by commissioned officers of the Prison Service. It was adorned with a gold and green knot.

Manufacture and distribution of both the Justice and Prison Service sabres ceased at the end of 1941, due to the exigencies of the war.

## RAILWAY SERVICE

During the imperial and Weimar eras, officers of the *Eisenbahnschutz* were provided with a gold-plated *Galadegen* for wear on dress occasions. Produced by Carl Eickhorn, it took the form of a slim rapier with a C-shaped knuckle bow and bore the traditional railway emblems of a winged wheel and a winged helmet on the shell guard and quillon block respectively. The hilt and scabbard fittings were ornately engraved with floral and linear motifs.

In 1935, further use of the pre-Nazi *Degen* was prohibited, and *Bahnschutz* officers were instructed to kit themselves out instead with one of the following two patterns of sabre already on the market:

- the Eickhorn Wrangel model, which had a gold oakleaf pommel and an open-winged eagle and swastika on the langet
- The Eickhorn Roon model, which had a gold oakleaf pommel and a plain langet upon which a suitable emblem could be engraved if desired

The wearing of these swords by Railway Guard officers ceased in April 1938, with the introduction of the *Bahnschutz* dagger.

It is noteworthy that the Eickhorn firm appears to have enjoyed sole manufacturing rights for all Railway Service edged weapons produced throughout the imperial, Weimar and Third Reich periods.

## MINING

The vital importance of the miner to the German economy was recognised long before the advent of the Third Reich, and as a reflection of this employees of the *Bergbau*, or mining industry, were provided with military-type black uniforms for wear at festive events. These outfits were based on the styles of dress used by miners and soldiers during the eighteenth and nineteenth centuries, with tall peakless shakos and long rows of silver or gold buttons on the tunic. When marching or in procession, ordinary miners traditionally carried axes with slender handles, while officials and musicians wore swords.

### MINERS' SABRES

Imperial Germany produced a wide variety of ornate miners' sabres, usually with grey sharkskin grips, which were often passed on from father to son and worn with pride across the generations. In 1933, in an effort to achieve some standardisation, the following three basic styles were formally approved for future use by *Bergbau* officials:

**136**. *An imperial or Weimar-era miner's sabre, with sharkskin grip, which was 'Nazified' after 1933 by the addition of a swastika to the crossed hammers emblem on the langet. Personalised variations such as this were commonplace amongst* Bergbau *officials.*

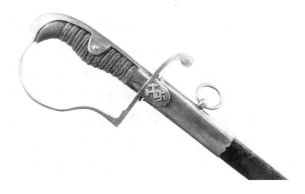

**137**. *A mining official in ceremonial uniform in 1935, with a short imperial-style* Bergbau Degen *hung vertically from a frog at his left side.*

- the Army-pattern *Einheitssäbel*, with black plastic grip, gold or silver fittings, and crossed hammers on the langet
- the Army-pattern lion-head sabre, with black plastic grip, gold or silver fittings, and crossed hammers on the langet
- a basket-hilted sabre, with black plastic grip, gold or silver fittings, and crossed hammers on the knuckle guard and scabbard

A number of variations on these themes were still permitted, including the incorporation of ornately etched blades bearing floral patterns and the miner's motto *Glück auf!* ('Good luck' or 'Keep safe'). The weapon was finished off with a gold bullion portepée knot having black and white striped decoration.

The *Bergbau-Säbel* was one of the few Third Reich edged weapons not to feature the eagle and swastika as part of its design.

## RIFLE ASSOCIATION

During the early years of the Third Reich, members of the *Deutscher Schützen Verband* wore a variety of swords, particularly the *Einheitssäbel* pattern, with their local rifle-club uniforms. Many of these individuals were First World War veterans, and simply continued sporting the edged weapons which they had carried in the Kaiser's army. The March 1936 issue of the cutlers' trade magazine *Die Klinge* illustrated a *Stichdegen*-type rapier which it specifically designated as a *Schützenwaffen*, or rifleman's weapon, but the accompanying text reinforced the fact that there were many manufacturers and models of shooting-club sword, and that selection was a matter for the purchaser.

Further wear of sabres and rapiers by members of the German Rifle Association was prohibited following the introduction of the *D.Sch.V.* cutlass in February 1939.

## PRESENTATION SWORDS

In addition to the range of regulation swords, rapiers and sabres worn by members of the various

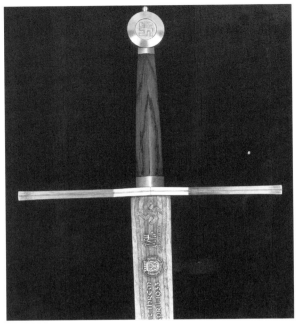

**139**. *The Adolf Hitler Solingen Sword of Honour. The city seal and* Oberbürgermeister*'s signature are highlighted on the ricasso of the Damascus steel blade, while the pommel reverse bears the inscription* Me Fecit Solingen *('Solingen Made Me').*

**138**. *Willy Liebel,* Oberbürgermeister *of Nuremberg, presenting Hitler with the Reich Sword in 1935.*

Nazi organisations, a small number of unofficial pieces were commissioned for special presentation purposes. The following few examples give an indication of the types involved.

### The Adolf Hitler Solingen Sword of Honour, 1933

On 3 April 1933, the *Oberbürgermeister* of Solingen, Dr Dotte, and his councillors voted to make Adolf Hitler an honorary citizen of their town. To commemorate the event, a massive Sword of Honour was commissioned under the auspices of the Solingen Industrial Trade School, with several factories working together to produce the

magnificent gift. Of typical *Schwert* design, with a cross-like hilt by Gottlieb Hammesfahr and an oak grip, the sword featured a maidenhair-pattern Damascus steel blade forged by Paul Müller. The obverse of the blade was etched by Karl Broch with the dedication 'With this sword, its noblest symbol, the city of Solingen confers honorary citizenship upon the *Führer* and Reich Chancellor, Adolf Hitler', while the reverse bore an excerpt from an old Germanic folk song extolling the virtues of the Bergischland, the mountainous region surrounding Solingen. The sword took five years to complete and was finally presented to Hitler in 1938.

### The Reich Sword, 1935

In 1935, *SA-Gruppenführer* Willy Liebel, *Oberbürgermeister* of Nuremberg, presented Hitler with a new sword of state known as the Reich Sword, based upon a twelfth-century original. The pommel bore the eagle of the Holy Roman Empire, and no fewer than 4,000 pearls and 18,000 gold pins were set into an ornate medieval diamond-shaped pattern on the scabbard. Hitler placed the sword on permanent public exhibition in the city museum at Aachen, Charlemagne's capital.

## THE SOLINGEN TRADE COMPETITION SWORD, 1940

The Solingen Industrial Trade School produced an exquisite sword as its entry for the city's trade competition in 1940. Designed by Paul Voss, the weapon's Damascus blade was forged by Paul Dinger. The hilt was of richly decorated oxidised steel, with an eagle and swastika on the langet, and the short crossguard bore the patriotic wartime dedication *Nur wer Stürmt hat Lebensrecht* ('Only those who struggle have the right to survive').

While a vast range of Nazi swords, sabres and rapiers was therefore approved, a number of influential uniformed paramilitary and state organisations were never authorised to wear such weapons. These groups included the following:

- the Corps of Political Leaders
- the National Socialist Motor Corps
- the National Political Education Institutes
- the Hitler Youth
- the German Air Sports Association
- the National Socialist Flying Corps
- the German Labour Front
- the government administration and Civil Service
- the Postal Protection Corps
- the National Labour Service
- the *Organisation Todt*
- the Technical Emergency Corps
- the Air Raid Protection League
- the Red Cross
- the Forestry Service
- the Hunting Association

Most Solingen blade-making firms employed professional artists to create potentially lucrative designs for them to consider, and throughout the 1933–42 period several companies submitted suggestions for new swords for the above and other organisations which were, for a variety of reasons, rejected by the authorities. Even towards the end of the war, when stringent production limitations had been imposed, sketches of new and innovative sword designs continued to flow from Solingen into the *Wehrmacht* and *NSDAP* headquarters in Berlin and Munich, in the forlorn hope that additional business might be created. It is perhaps significant that Hitler himself is never known to have carried a sword of any type, indicative of his deep-rooted personal identification with the common man rather than the officer class.

# PART THREE

# BAYONETS

While dress daggers and swords were arguably the most splendid of Nazi edged weapons, the humble bayonet was certainly the most practical and prolific. During the Second World War, in excess of 20 million examples were produced, with around 75 per cent of these being destined for the combat troops who were engaged in constant life-and-death struggles in the furthest reaches of Hitler's empire. If daggers and swords could talk, they might relate only tales of parades and pageants. Bayonets, on the other hand, would boast of campaigns and conquests, of heroism and hardship, and of death, destruction and defeat. They may have been the 'poor relations' in the Third Reich's family of blades, but they were expected, like their low-ranking wearers, to perform that family's most basic role – killing the enemy.

The word 'bayonet' derives from the French town of Bayonne, where this type of weapon originated, and technically speaking should refer only to a dagger or knife which has been made so as to fit over the muzzle of a firearm. The equivalent German term *Seitengewehr*, or 'sidearm', covers not only the bayonet *per se* but also its ceremonial variants, which were never intended for front-line

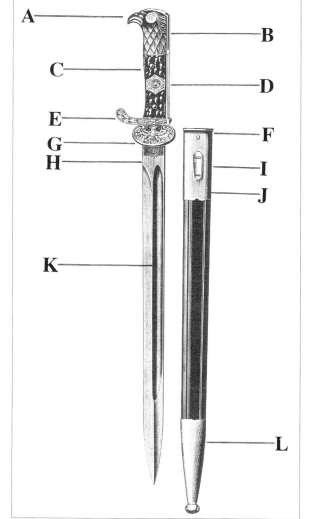

**140**. *Bayonet Nomenclature (Police M.26 clamshell bayonet with Weimar insignia as example):*
*A – Pommel/securing button; B – Rifle slot/mortise;*
*C – Hilt/grip plate; D – Grip insignia;*
*E – Quillon/crossguard; F – Throat;*
*G – Shell guard/clamshell; H – Ricasso; I – Frog stud;*
*J – Locket; K – Fuller/blood groove; L – Chape*

use and consequently were not provided with functional rifle slots.

Nazi *Seitengewehre* fell into the following distinct categories:

- standard service bayonets, for combat use
- dress bayonets, produced in a wide range of styles and qualities, for parade and on-duty wear by enlisted men and junior NCOs

Both were commonly adorned with a distinctive type of knot known as the *Troddel*, or 'tassel', which was brightly and elaborately coloured according to the bearer's company and battalion.

This chapter describes a representative selection of the bayonets which were officially authorised and used by the *Wehrmacht* and other Third Reich organisations. However, it should be borne in mind that it was not uncommon for contemporary bayonet owners to affix unit badges or formation insignia to the grips of their privately purchased weapons in an effort to personalise them, a practice which gave rise to an almost infinite variety of unofficial pieces surviving the war.

# ARMED FORCES

## *Wehrmacht* Service Bayonets

The standard-issue combat bayonet for the Army, Navy, Air Force, *Waffen-SS* and all other front-line forces of Hitler's Germany was the 38 cm

**141**. *It is hard to believe that these baby-faced soldiers shown 'monkeying around' in their barracks at the end of 1934 are members of Hitler's elite* SS *bodyguard regiment, the* Leibstandarte. *The training NCO at upper left, in peaked cap, is less amused than his charges. At least fifteen M.84/98 bayonets are in evidence. The trooper in the centre of the photograph, with both hands raised, seems to be the focus of attention and appears to have a bayonet sticking out of his chest!*

**142.** *Soldiers of the* Leibstandarte-SS *'Adolf Hitler' parading with M.84/98 bayonets fixed, 1940.*

*Dienstseitengewehr* M.84/98. This ubiquitous weapon was commonly referred to as the Mauser bayonet since it was designed to be affixed to the Mauser *Gewehr* 98, which had been adopted as the imperial Army's principal bolt-action rifle on 5 April 1898 and continued in use throughout the First and Second World Wars in the shortened guises of the Karabine 98a and Kar. 98k, respectively.

The M.84/98 bayonet was a substantial item constructed to withstand sustained battlefield use. It had a heavy-duty blued tempered steel blade with a double edged point, a blood fuller on each side and a flat upper edge. The non-reflective steel hilt featured a rifle slot or mortise with an associated spring-loaded securing button at the pommel end, while the grip plates could be of brown or black

bakelite or wood, and were held in place by two countersunk screws and slotted nuts. A ventilation hole was cut into the portion of the grip plates adjacent to the short quillonless crossguard, to allow any moisture trapped under the grips to escape and so avoid possible internal damage from rust.

The M.84/98 bayonet scabbard was again made of matt blued steel, seamlessly rolled and fitted with an integral ball end at the chape for added strength. It had an elongated stud welded to the front for looping through the blackened leather suspension frog. A so-called *Seitengewehrtasche für Berittene*, or 'bayonet frog for mounted personnel', which bore an additional 2 cm wide leather strap at the top to secure the grip of the bayonet and prevent it from swinging, was ordered to be issued to all troops, both cavalry and non-cavalry, from 25 January 1939. However, this direction was not rigidly adhered to. Many simplified versions of the regulation frog, and webbing examples for tropical use, emerged during the Second World War.

Each one of the 15 million or so M.84/98 bayonets produced between 1933 and 1945 had to pass rigorous examination by military armourers, whose inspection, or *Waffenamt*, stamps were thereafter applied to one or more of a number of locations, including the pommel, crossguard and blade. Scabbards and frogs were similarly scrutinised and marked. The *Waffenamt* stamps on bayonets manufactured prior to May 1934 featured the Weimar Republic-style eagle, without swastika, while later examples bore the Nazi version.

Before the outbreak of the Second World War, each bayonet and scabbard was also normally stamped with the maker's full name and town of origin, and the date of manufacture. From 1940, however, these marks were replaced by one of a range of three-letter codes allocated to military equipment suppliers for security reasons. Many hundreds of these codes were eventually devised, and the following are known to have been applicable specifically to the larger Solingen edged weapons firms, who were contracted to produce service bayonets:

asw – E. & F. Hörster
clc – Richard Herder
cof – Carl Eickhorn
cqh – Clemen & Jung
crs – Paul Weyersberg
cul – Ernst Pack & Söhne
edg – J.A. Henckels
ffc – Friedrich Herder
fnj – Alexander Coppel
fze – F.W. Höller

Such codes were usually accompanied by the final two digits of the year of production, for example 'asw 43'.

Bayonet blades were further marked with issue accountability numbers, centrally allocated in batches and cross-referenced to the records of the recipient soldier. Such numbers ran from 1 to 9999. Once the 10,000th bayonet had been reached, a letter was used for the next series, e.g. 1a to 9999a. This was continued through the alphabet. On the first day of January each year, the numbering system started all over again with 1. Naval bayonets usually bore the *Kriegsmarine* property symbol of an eagle over the letter 'M', while some Air Force and Railway Service examples were stamped '*RLM*' (*Reichsluftfahrtministerium*, or Air Ministry) and '*RBD*' (*Reichsbahndirektion*, or National Railway Directorate) respectively. Front-line auxiliary formations such as the *Organisation Todt*, the *Reichsarbeitsdienst* and, late in the war, the *Volkssturm*, are also known to have marked a few *Seitengewehre* with their own distinctive property logos.

The M.84/98 service bayonet was the true 'workhorse' amongst *Wehrmacht* edged weapons, being used for a multitude of purposes from probing for hidden mines and cutting emergency bandages to opening tin cans and slicing loaves of bread. It is a testimony to the excellence of its construction that very few examples ever required replacement or field repair, other than sharpening, with the vast majority giving sterling service to their owners throughout the war.

To supplement the regulation *Seitengewehr*, which was always expensive to produce, considerable quantities of captured Polish, French, Belgian, Russian and other bayonets were pressed into service during 1940–43, along with their accompanying firearms. Obsolete German models including the M.98/05 bayonet, similar in design to the M.84/98 but with the addition of a single long curved quillon, were also reissued to rear-echelon troops and second-rate foreign volunteer units, particularly in 1944. For reasons of both economy and morale, however, Nazi markings on these pieces were kept to a minimum and, indeed, there was often no stamp at all which would indicate Third Reich approval and use.

## *Wehrmacht* Dress Bayonets

Junior NCOs, below warrant officer grade, and enlisted men of the *Wehrmacht* and *SS-VT* had the option of privately purchasing dress bayonets known as *Extra-Seitengewehre* for wear when walking out on duty. These usually took the form of the M.98 bayonet, with its short curved quillon, and were produced in a 35 cm length for NCOs and 40 cm for men. Quillonless variants fashioned after the *Dienstseitengewehr* M.84/98 were also available for enlisted ranks only, and in this context were often referred to as 'pioneer models'.

---

**143**. *A* Wehrmacht *enlisted man's M.98 dress bayonet with the standard commemorative dedication* Zur Erinnerung an meine Dienstzeit *etched on the blade.*

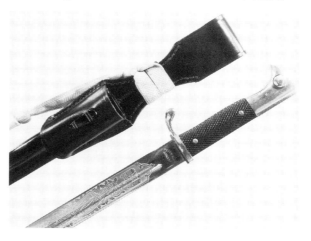

The dress bayonet had a rifle mortise but could not be fitted to a firearm. While its spring-loaded pommel button was usually functional, the accompanying slot was deliberately too small for attachment to the Kar. 98k and was often filled in with a decorative piece of coloured woollen cloth appropriate to the service of the wearer.

The typical *Extra-Seitengewehr* had silver-toned fittings, a chequered black plastic grip, a polished steel blade and a gloss black stove-enamelled steel scabbard. The frog was in black patent leather, or brown leather for the *Luftwaffe*, and a *Troddel* finished the sidearm off. However, numerous variations on this general theme were produced simultaneously, to maximise choice for the purchaser, so that the following features might be encountered:

- nickel-plated steel hilt fittings
- nickel-plated zinc hilt fittings
- chrome-plated steel hilt fittings
- chrome-plated zinc hilt fittings
- nickel-plated blades
- chrome-plated blades
- genuine staghorn grips
- imitation staghorn grips
- wooden grips

---

**144**. *The etched blade of this* Wehrmacht *NCO's dress bayonet has a sawtooth edge, traditionally associated with engineer or assault units.*

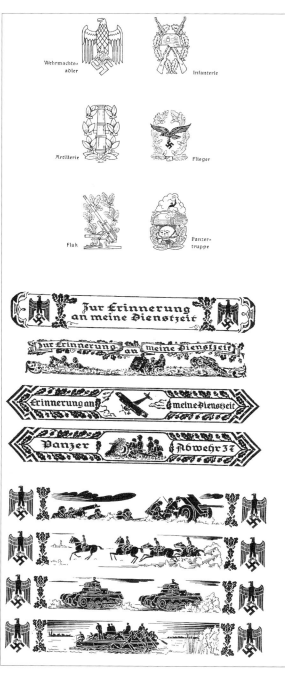

**145**. *A selection of the many insignia and decorative designs and dedications which could be etched on to* Wehrmacht *dress bayonets at extra cost.*

*146. A bayonet blade-etching produced for members and veterans of the 66th Infantry Regiment, based at Magdeburg.*

Prices varied accordingly with, for example, a standard nickel-plated steel bayonet retailing at 3.80 Reichsmarks and one with staghorn grips selling at 4.20 Reichsmarks. Wholesale distributors placing bulk orders with makers were afforded discounts at the following rates:

- 20–50 bayonets . . . . . . . . . 5% discount
- 51–100 bayonets . . . . . . . .10% discount
- 101–200 bayonets . . . . . . . .15% discount
- Over 200 bayonets . . . . . . .20% discount

During the 1935–9 period, numerous commemorative versions of the dress bayonet were also made available, not for wearing but for displaying as mementoes of conscript service. These could be bought by the soldiers themselves, but were more often than not purchased by proud parents for gifting to a son on the completion of his two-year term of compulsory military duty. The blades of such pieces were frequently etched with the standard dedication *Zur Erinnerung an meine Dienstzeit* ('To commemorate my national service'), to which could be added the owner's name, unit, appropriate dates and symbolic decoration. In addition, pommel embellishment was available in the form of etched insignia representing the infantry, artillery and other *Wehrmacht* branches. Prices were set according to the individual specification of the bayonet ordered, and the following quotes by the WKC firm in 1938 give an indication of costs involved:

- to etch a standard dedication on the obverse side of a blade – 0.60 Reichsmarks
- to etch both sides of a blade – 1.40 Reichsmarks
- to highlight an etched dedication in gold-plate – 0.30 Reichsmarks extra per letter
- to etch an insignia on a pommel – 1.00 Reichsmark

A nickel-plated bayonet with basic single-etched blade would therefore have cost around 4.40 Reichsmarks, compared with one with staghorn grips, pommel decoration and a double-etched gold-plated blade, which would retail at 20.10 Reichsmarks. No bulk discount could be applied to the wholesale ordering of bayonets with etched blades.

The supply of dress and commemorative bayonets was therefore a very lucrative business and represented a major source of revenue for the Solingen edged weapon manufacturing companies. Competition between these firms was keen, resulting in an almost limitless variety of bayonet embellishments being marketed to suit all pockets and tastes.

A third category of *Extra* bayonet was the *Ehrenseitengewehr*, presented as a unit prize or trophy for marksmanship, success in annual inter-service competitions, or similar achievement. Bayonets of Honour typically featured eagle-head pommels, oakleaf decoration to the hilt, staghorn grips and suitably etched or engraved blades. They were purchased by companies, battalions and regiments and were either gifted permanently to the recipients or, in the case of sports and other contests, loaned on a temporary basis for a specified period of time.

Around 5 million *Wehrmacht* dress bayonets were produced between 1935 and 1941. Not being official armed forces issue items, they were never subject to armoury inspection so were not *Waffenamt* stamped, coded or serial numbered in the manner of the *Dienstseitengewehr*. Most bore only the maker's mark, and occasionally the distributor's or retailer's stamp, on the blade. A small number were later adopted as issue bayonets by the *Postschutz* and other uniformed non-combatant civil organisations, and these were embossed with appropriate property designations such as '*DRP*' (*Deutsche Reichspost*).

## WEHRMACHT BAYONET KNOTS

Peacetime regulations dictated that a bayonet knot or *Troddel* was to be attached to the frog of all *Seitengewehre*, except when worn with field dress.

Army NCOs' *Troddeln* were universally dark green in colour, with silver thread patterning. Those for enlisted men, however, had their component parts coloured in a variety of ways according to the following arrangement:

- The strap was grey, except for regimental and battalion staff and troops of the 13th and 14th companies, whose straps were dark green.
- The slide was in company colours.
- The stem was in battalion colours.
- The crown was in company colours.
- The tassel was in grey or dark green, to match the strap.

**147**. *An Army Bayonet of Honour, with eagle-head pommel, manufactured by Paul Weyersberg and presented as a prize for marksmanship during 1937–8.*

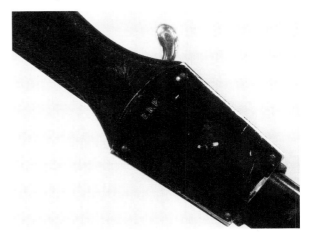

**148**. *A standard military dress bayonet frog stamped with the 'DRP' ('Deutsche Reichspost') property mark, denoting issue to a member of the* Postschutz.

## Police Bayonets, 1920–33

In 1920, plain Army surplus M.98/05 service bayonets, with their single long curved quillons, were issued as symbols of authority for everyday wear by members of the new provincial police forces set up in terms of the Weimar constitution. A full-dress version of this sidearm pattern, known as the *Polizeiseitengewehr Sonderausführung*, or *PSS*, was created at the same time for private purchase and use on ceremonial occasions. It was entirely nickel-plated with oakleaf decoration to the hilt and crossguard. The Weimar Police emblem, comprising a six-sided star with a static and weaponless republican eagle in the centre, was affixed to the staghorn grip, while the nickelled blade was modified to incorporate a spear point and was contained in a plain blued steel scabbard. Worn with a knot in provincial colours, the *PSS* was the first in what was soon to become a long line of Police dress bayonet variations.

In 1926, the state of Prussia inaugurated its own series of ornate Police bayonets based on the M.98 *Seitengewehr* with short curved quillon. Unlike the *PSS*, however, these *Polizei-Dienstseitengewehre* were now issued to all non-commissioned Police personnel for everyday service use. Blade lengths ranged from an unwieldy 57 cm for *Unterwachtmeister* (the lowest police rank) to 40 cm for senior NCOs up to and including *Hauptwachtmeister*. Each weapon had a decorative pommel cast in the form of an eagle's head, with or without rifle mortise, and a shell guard or 'clamshell' bearing the Weimar eagle. Once again, the republican Police star adorned the staghorn grip. The M.26 scabbard could be in black leather for municipal Police officers or brown leather for their rural colleagues, and had nickel-plated mounts. In 1933, Police bayonet blade lengths were standardised at a less cumbersome 33 cm.

All issued bayonets were stamped on the hilt and upper scabbard mount with matching accountability markings consisting of letters, Roman numerals and Arabic numerals. Letters indicated the Police departments or administrative districts involved, while Roman numerals related to particular Police stations within departments or districts and Arabic numerals represented individual weapon serial numbers. So, for example, the marking 'SB.II.41' would denote '*Schutzpolizei* Berlin, Police station no. II, bayonet no.41'. 'LS.III.27' would indicate '*Landjägerei* Schneidemühl, Police station no. III, bayonet no.27'. The following table lists all the abbreviations known to have been used on Prussian police bayonets. However, it should be noted that this list is not exhaustive since Bavarian, Saxon and other state forces which were later incorporated into the *Deutsche Polizei* had their own equivalent series of abbreviations. Unfortunately, no definitive source for these is known to have survived.

| Abbreviation | Police Department or Administrative District |
|---|---|
| A | Aurich District |
| Al | Allenstein District |
| An | Aachen District |
| Ar | Arnsberg District |
| B | Berlin District |
| Br | Breslau District |
| D | Düsseldorf District |
| E | Erfurt District |
| F | Frankfurt-on-Oder District |
| G | Gumbinnen District |
| H | Hanover District |
| Hi | Hildesheim District |
| Hp | Central Police School |
| K | *Kriminalpolizei* (Criminal Investigation Department) |
| K | Köslin District |
| Ka | Kassel District |
| Kg | Königsberg District |
| Kö | Köln District |
| Kz | Koblenz District |
| L | *Landjägerei* (Rural Police, later the *Gendarmerie*) |
| Lg | Lüneburg District |
| Li | Liegnitz District |
| LSAl | Allenstein Rural Police School |
| LST | Trier Rural Police School |
| M | Münster District |

| | |
|---|---|
| Me | Merseburg District |
| Mg | Magdeburg District |
| Mi | Minden District |
| O | Osnabrück District |
| Op | Oppeln District |
| P | Potsdam District |
| PB | Bonn Police School |
| PBd | Brandenburg Police School |
| PBg | Burg Police School |
| PFr | Frankenstein Police School |
| PHi | Hildesheim Police School |
| PK | Kiel Police School |
| PL | Police Physical Training School |
| Pl | Central Police Academy |
| PM | Münster Police School |
| PMd | Minden Police School |
| PS | Sensburg Police School |
| PT | Treptow Police School |
| PTV | Police Technical and Communications School |
| RhP | Rhine River Police |
| S | Schneidemühl District |
| S | *Schutzpolizei* (Municipal Police) |
| Sch | Schleswig District |
| Sd | Stralsund District |
| Si | Sigmaringen District |
| St | Stettin District |
| Sta | Stade District |
| T | Trier District |
| W | Wiesbaden District |
| Wpr | West Prussia District |

## THE *FELDJÄGERKORPS* 1933 BAYONET

In October 1933 troops of Göring's Nazi auxiliary police force, the *Feldjägerkorps Preussen*, were issued with a special version of the M.26 Prussian Police bayonet distinguished solely by the incorporation into its grip of the distinctive Berlin Interior Ministry and *Feldjägerkorps* emblem. This again comprised a six-pointed star with an eagle in the centre, but in this case the bird was of the imperial Prussian flighted variety, clutching a sword and lightning bolts in its talons, and had a swastika on

its chest. The weapon was worn with a black, silver and gold knot. On 1 April 1935 the *FJK* was wholly absorbed into the *Schutzpolizei* and adopted regular Police bayonets, at which time further wear of the *Feldjägerkorps-Seitengewehr* ceased.

## POLICE BAYONETS, 1933–45

After Hitler's assumption of power and the formation in 1934 of the *Deutsche Polizei*, the M.26 Prussian-pattern bayonet, minus its shell guard, was

**152**. *A Police M.36 service bayonet. The reduced and ill-proportioned grip plates indicate that this 33 cm-bladed piece has been produced by converting an older Prussian 40 cm/57 cm-bladed Police bayonet. Note that the* Unterwachtmeister *knot has been attached incorrectly to the hilt rather than to the suspension frog.*

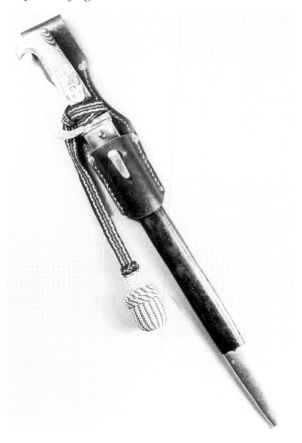

*Wachtmeister*, i.e. personnel below warrant-officer grade who were not entitled to sport the new *Polizei-Unterführerdegen*.

M.36 Police service bayonets made to Nazi specification were fairly expensive, retailing at 13.50 Reichsmarks with a rifle mortise or 11 without the mortise. Consequently, throughout the 1937–9 period, the majority of old Prussian Police-issue bayonets were recalled to local armouries for reasons of economy, to have their obsolete republican clamshells removed, Nazi insignia affixed and their blades shortened to the regulation 33 cm. This latter operation necessitated a simultaneous

**154**. *A Police M.36 dress bayonet by WKC, with 25 cm blade.*

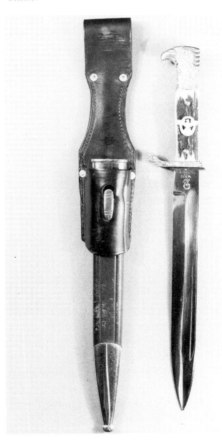

**153**. *Accountability stamps on a Police bayonet issued by the Berlin* Schutzpolizei. *This particular example does not indicate the station of origin, but simply gives the weapon inventory serial no. '10748'.*

approved for use nationally. A newly designed police insignia, featuring the small M.29/34 *NSDAP* eagle surrounded by laurel leaves, was soon amended to depict the larger state eagle surmounting a wreath of oakleaves. This revised emblem duly replaced the Police star badge on bayonet grips from 1936, coinciding with Himmler's appointment as Chief of Police and the introduction of a national Police uniform. The wearing of service bayonets, or *Polizei-Dienstseitengewehre*, was now restricted to *Unterwachtmeister* and reduction in hilt size, for the sake of balance and overall appearance.

A shorter version of the M.36 *Polizei-Dienstseitengewehr*, with a 25 cm blade, was authorised for private purchase only and tended to be of lighter weight and higher quality than the regulation-issue bayonet. This so-called *Polizei-Extraseitengewehr* did not bear issue stampings of any kind and could be bought direct from retailers for a price of 9.50 Reichsmarks upon provision of the buyer's Police identity documentation. Moreover, the M.20 *PSS* bayonet continued to be produced in small numbers until 1940 as an alternative form of private-purchase sidearm. Various commemorative bayonets and Bayonets of Honour with etched blades, similar to those of the *Wehrmacht*, were also made available for presentation to police sports competition winners, retiring officers, etc. on a unit basis.

The regulation Police bayonet knot comprised a black leather or cloth strap with red and aluminium thread decoration and a silver acorn. Knots for *Unterwachtmeister* featured green vertical stripes to their acorns.

The manufacture of Police bayonets ceased in 1941, by which time it had become normal practice for appropriate Police personnel to be issued with, or to buy, much cheaper standard *Wehrmacht* dress bayonets for duty wear. During 1943, the carrying of pistols became almost universal due to the deteriorating war situation and Police bayonets were subsequently no longer in general evidence on the streets of Germany, except on rare festive occasions.

# FIRE BRIGADE

From 27 May 1936, Fire Brigade personnel below the rank of *Hauptbrandmeister* had their dress axe superseded by the so-called *Faschinenmesser*, or machete, an unadorned bayonet-like sidearm measuring 35 cm for NCOs and 40 cm for enlisted men. Available through private purchase only, its design was based on the standard M.98-pattern military dress bayonet with nickel fittings and black bakelite grip, but the rifle mortise and stud button

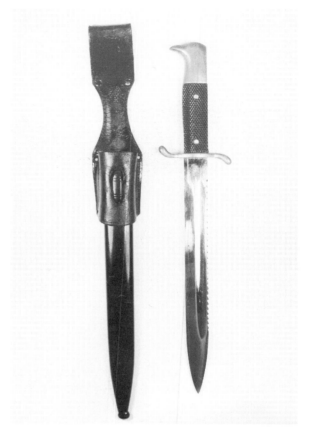

**155**. *A Fire Brigade Faschinenmesser, with sawtooth back edge to the blade.*

were absent from the hilt and the crossguard was lengthened on both sides to give a distinctive recurved S-shape. The steel blade could be obtained with a plain, nickel-plated or chrome-plated finish, and was manufactured either with or without a *Sägerücken* or sawtooth back edge. The *Faschinenmesser* was contained in a *Wehrmacht*-style black stove-enamelled steel scabbard and was suspended from a black leather frog. Its accompanying knot was similar to that of the Police, but with carmine red vertical stripes to the acorn for those of the lowest ranks of *Feuerwehrmann* and *Brandmeister*.

The 40 cm Fire Brigade bayonet retailed at 3.50

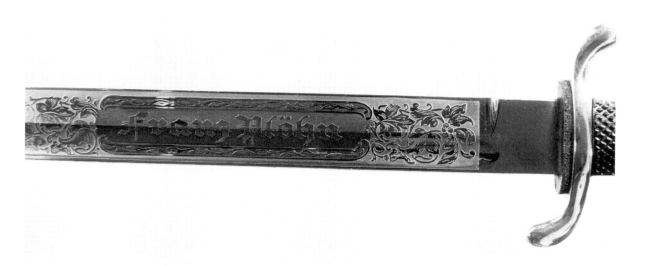

*156. The blade of this Fire Brigade sidearm is etched with the owner's name, 'Franz Plöhn'.*

Reichsmarks, while the 35 cm version sold for 3.30. Sawtooth blades cost an additional 0.60 Reichsmarks. A few examples were etched with commemorative and other celebratory inscriptions, but these were rare since the old custom of presenting suitably engraved axes to firemen as retirement gifts or tokens of appreciation continued well into the Second World War.

## CUSTOMS SERVICE

During the early years of the Third Reich, junior customs officials were equipped with plain M.98 bayonets which they wore on regular duty tours. In 1936, however, *Landzoll* and *Wasserzoll* personnel below the ranks of *Oberzollsekretär* and *Maschinenbetriebsleiter* respectively were authorised to carry a more ornate dress sidearm with the service uniform. This new Customs Service bayonet was identical to the 33 cm M.36 *Polizei-Dienstseitengewehr*, but without the grip insignia. Fittings were silver-plated for Land Customs staff and gold-plated for Water Customs, and the scabbard and hilt normally bore the stamped property mark '*RFV*' (*Reichsfinanzverwaltung,* or National Finance Administration), accompanied by an accountability serial number. A shorter version with a 25 cm blade, akin to the *Polizei-Extraseitengewehr*, was also made available for private purchase.

The production of Customs Service bayonets was halted in 1941, after which *Zollgrenzschutz* personnel were issued with, or bought, standard *Wehrmacht* dress bayonets for daily wear.

## AIR RAID PROTECTION LEAGUE

In 1936, a dress bayonet was made available for private purchase by full-time NCOs and subordinate personnel of the *Reichsluftschutzbund*. Eligibility was restricted to those staff between the ranks of *Obertruppmeister* and *Truppmann* who had not been presented with the more prestigious *RLB-Messer*, and only a very few individuals appear to have availed themselves of the opportunity to buy the new weapon.

The *Luftschutz-Seitengewehr* took the form of a standard *Wehrmacht* 35 cm dress bayonet with the addition to the grip of the organisation's insignia, comprising an eight-pointed sunburst surmounted by the initials '*RLB*' over a swastika. From mid-1938, this badge was replaced by the revised *RLB*

**157**. *An* RLB *bayonet with staghorn grips.*

**158**. *An NSKOV bayonet, by E. & F. Hörster.*

emblem of a large black swastika set upon a sunburst.

Any wearer of the *Luftschutz-Seitengewehr* who was subsequently awarded the *RLB-Messer* was obliged to discard the bayonet in favour of the knife.

Manufacture of this rare bayonet was discontinued at the outbreak of the Second World War.

## VETERANS' ASSOCIATIONS

After the First World War and the banning of *Freikorps* formations in 1921, the growth of the already plentiful military veterans' organisations throughout Germany was enormous, with dozens of new leagues of former soldiers being established. Some, such as the *Frontkriegerbund München* and the *Landesverband Bayern*, based

their membership criteria on residence locality. Others, like the *Ehrenbund der Verdun und Argonnenkämpfer* and the *Waffenring der Deutschen Schweren Artillerie*, accepted only those who had seen action in nominated campaigns or with particular branches of the armed forces. An ex-artilleryman resident in Munich who had fought at Verdun would be eligible to join all of the above formations, and in this way multiple membership of associations became commonplace during the 1920s and early 1930s.

The *Stahlhelm*, or Steel Helmet, League acted as an umbrella group, attracting all former front-line soldiers into one national organisation. Its development was phenomenal, reaching a membership of 1 million in 1930, by which time it included youth divisions, female sections and even an air squadron. Theoretically a welfare and mutual assistance body for 'old comrades', the *Stahlhelm*'s real objective was the downfall of the Weimar Republic and the installation of a right-wing regime with traditional militaristic values. The Steel Helmet leader, Franz Seldte, became a devoted supporter of Hitler, and after the Nazi assumption of power in January 1933 he was appointed Reich Minister of Labour. The best of his members were subsequently summoned to act as Police auxiliaries for the duration of the anti-Communist purges, and many went on to join the *SA* and *SS*. By a decree of 28 March 1934, the already close ties with the *NSDAP* were formalised when the *Stahlhelm* League was affiliated to the party and renamed the *Nationalsozialistisches Deutscher Frontkämpferbund Stahlhelm*, or *NSDFBSt*.

On 7 November 1935, its aims achieved, the *NSDFBSt* was honourably dissolved by Hitler. Former *Stahlhelm* men who had not already joined a Nazi paramilitary formation were now ordered to enrol in yet another sympathetic group, the *Kyffhäuserbund der Deutschen Landkriegerverbände*, named after the celebrated imperial Lighthouse war memorial. The *Kyffhäuserbund* had gradually united numerous provincial military and *Freikorps* old comrades' leagues and societies, and on 4 March 1938 it too was absorbed into the *NSDAP* and renamed the *Nationalsozialistisches Reichskriegerbund*, or *NSRKB*, becoming the only permitted veterans' association in Germany. Led by *SS-Gruppenführer* Reinhard, the *NSRKB* thereafter worked in close co-operation with the Nationalsozialistisches *Kriegsopferversorgung*, or *NSKOV*, the Nazi War Victims' Support Service, which looked after the welfare of casualties of the First and Second World Wars, as well as that of those party stalwarts who had suffered in street fighting with Communists during Hitler's struggle for power.

Many of these veterans' associations were fully or partly uniformed, and encouraged their members to wear edged weapons on parade. Bayonets were a favourite sidearm in this respect, often being specially purchased and adapted by the owners themselves to incorporate some form of organisational insignia. Several variants duly emerged, including bayonets of the *Polizei-Dienstseitengewehr* pattern modified so as to include the emblems of the *Stahlhelm* or *NSKOV* on the staghorn grips. Such items were wholly unofficial, but were tolerated by the authorities as their proud wearers were generally elderly or disabled individuals who had demonstrated considerable self-sacrifice in the service of their country.

---

**159**. *A customised M.98 dress bayonet, with a* Wehrmacht *eagle insignia attached to the grip. Also of interest is the unusual fire helmet trademark stamped into the blade ricasso.*

The vast majority of bayonets officially authorised and mass produced during the Third Reich therefore related to only three groups of organisations, namely the armed forces, the Police and the Fire Brigade. However, while the range of *Seitengewehre* was less extensive than that of daggers and swords, there was far more scope for personalisation and individuality where privately purchased bayonets were concerned. Relatively cheap etching and engraving meant that even low-ranking buyers could easily have their pieces made to order at will, and those owners who intended keeping their weapons as mementoes rather than for actual wear regularly customised them by affixing unit emblems, etc. This resulted in an almost limitless variety of bayonets being produced. Some of those which have been encountered include:

- standard M.98 military dress bayonets with the *Wehrmacht* eagle, *SS* runes or a death's head on the grip

- an M.26-pattern Prussian Police bayonet with the addition of the badge of the pre-Nazi German Hunting Association on the hilt
- an *HJ-Wachgefolgschaft* bayonet with the Hitler Youth diamond insignia replaced by that of the National Socialist Students League

Various doomed prototypes were also put together, such as the excessively ugly one proposed for the German Red Cross, which featured an M.98 dress bayonet grip bearing a red cross, a blade like that of the *DRK-Haumesser* and a large swastika on the crossguard.

As with daggers and swords, bayonet manufacture mirrored the fortunes of the Reich. From 1942, Solingen's forges were devoted almost entirely to output of the M.84/98 *Seitengewehr* and to production of what was perhaps the most practical of all Nazi edged weapons – the fighting knife.

# PART FOUR

# KNIVES

Several distinctly different types of Nazi military knife were produced, all of them falling into one of the following two groupings, depending on their intended purpose:

- fighting knives, for use as weapons during close-quarter hand-to-hand combat
- utility knives, for use as practical tools to assist servicemen in their daily chores at the front

Each of these categories is best described separately, although it has to be recognised that there was inevitably a great deal of flexibility in how individual knives were actually used by those who carried them.

## FIGHTING KNIVES

### FIGHTING KNIVES, 1914–18

Fighting knives, or *Nahkampfmesser*, originated early in the First World War when the close-combat capability of regulation service bayonets was soon found to be poor because of their excessive size and weight. The M.84/98, M.98, M.98/05, M.14 and a host of *ersatz* or substitute *Seitengewehre* all had single-edged blades ranging in length from 25 cm to 57 cm which regularly broke during determined thrusts. It became apparent that the best type of weapon for close-quarter fighting would be a sturdy dagger with a double-edged blade about 15 cm long and a short crossguard. The need for such a knife was accelerated by the advent of the storm troops.

Once the stalemate of trench warfare had set in at the end of 1914, Germany was quick to realise the advantage of developing elite units of hand-picked infantrymen to act as assault parties and trench raiders. Early in 1915 Major Eugen Kaslow, a pioneer officer, was tasked with evaluating experimental steel helmets, body armour and a new light cannon. To do so, he put together a small assault detachment which came to be known as *Sturmabteilung Kaslow*. Under his leadership and that of his successor, *Hauptmann* Willi Rohr, the *Sturmabteilung* evolved new tactics to break into an enemy trench system. Combat operations in the Vosges Mountains that autumn suggested that these ideas were sound, and in January 1916 *Sturmabteilung* Rohr was duly transferred to Verdun. At the time, the detachment comprised three-man teams called *Stosstruppe*, or shock troops, whose method of attack involved storming a trench in flank. The first of the trio was armed with a sharpened entrenching tool and a shield made from a machine-gun mounting. He was followed by the second man carrying haversacks full of short-fused stick grenades and the third soldier armed with a knife or club. The *Stosstrupp* technique proved so successful that a number of *Sturmkompanien*, or assault companies, were soon formed and attached to divisions on a permanent basis. By 1918, most German armies on the western front had expanded units known as *Sturmbataillonen*, or assault battalions, and the storm troops, as they became known to their British adversaries, were accorded the status of romantic heroes by the German popular press.

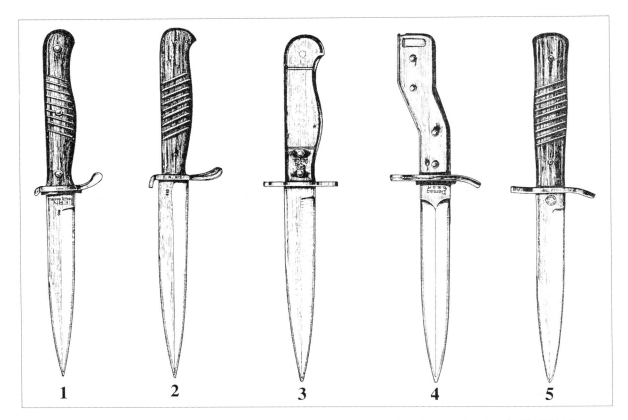

**160**. *Five typical First World War fighting knives, some of which saw renewed use between 1939 and 1945. 1 – ERN-made knife, dated 1915; 2 – Knife stamped '2.K.103' (2nd Company, 103rd Infantry Regiment), c. 1915; 3 – Steel-hilted knife, c. 1915; 4 – DEMAG-made all-steel cranked knife, designed to fit as a bayonet on both the G.98 and K.98 rifles, 1915; 5 – DEMAG-made knife, dated 1916*

Unlike ordinary infantrymen, they spent little time skulking in filthy trenches. Instead, they attacked suddenly then returned to base with the inevitable cache of prisoners. Needless to say, casualties and rates of mortality were high among these soldiers, the earliest exponents of *Blitzkrieg*. Officially, special insignia for the storm troops was frowned upon, but many varieties of locally adopted badges were worn, typically featuring death's heads, steel helmets, hand grenades – and knives.

So it was that during the first half of 1915 several middle-range German cutlers, such as Deutsche Maschinenfabrik of Duisburg and ERN Rasier-messerfabrik of Solingen, began to manufacture trench daggers to supplement the variety of short hunting knives which had already made their way to the front as private purchases. A wide diversity of designs duly emerged, most being conventional general-purpose cut-and-thrust weapons with double-edged blades and crossguards to catch an opponent's dagger. In 1917, officers serving in battle areas were finally ordered to discard their swords and wear trench knives instead, giving rise to the ludicrous situation whereby their elaborate gold and silver bullion portepée knots flourished as rank symbols on the plain fighting daggers produced for common soldiers. Attempts to devise a more ornate combat knife specifically for officers led to the production of a dwarf version of the M.98 bayonet with chequered grips in wood or horn, which developed through the Weimar Republic into the *HJ-Fahrtenmesser* and the plastic-gripped side-arms of the Third Reich.

A small but representative selection of the scores of variant fighting knives and battlefield conversions used between 1914 and 1918 is described below. Many of these were subsequently passed from father to son and later saw active service again during the Second World War, so they merit some degree of coverage when considering the edged weapons of Hitler's Germany.

### TYPE 1

| | |
|---|---|
| Manufactured: | 1914 |
| Total Length: | 270 mm |
| Blade Length: | 138 mm |
| Hilt: | Stag's hoof with S-shaped quillons |
| Markings: | None |
| Scabbard: | Black leather with belt loop |
| Remarks: | Private-purchase hunting knife |

### TYPE 2

| | |
|---|---|
| Manufactured: | 1914 |
| Total Length: | 269 mm |
| Blade Length: | 137 mm |
| Hilt: | Staghorn with S-shaped quillons |
| Markings: | D. Peres, Solingen and ale cask trademark |
| Scabbard: | Brown leather with belt loop |
| Remarks: | Private-purchase hunting knife |

### TYPE 3

| | |
|---|---|
| Manufactured: | 1914 |
| Total Length: | 321 mm |
| Blade Length: | 197 mm |
| Hilt: | Pressed steel, imitating staghorn, with S-shaped quillons |
| Markings: | Weyersberg, Kirschbaum & Cie. |
| Scabbard: | Blackened steel with bayonet frog |
| Remarks: | Private-purchase hunting knife |

### TYPE 4

| | |
|---|---|
| Manufactured: | 1914 |
| Total Length: | 268 mm |
| Blade Length: | 145 mm |
| Hilt: | Wood with short straight quillons |
| Markings: | King's head and knight's helmet of WKC |
| Scabbard: | Blackened steel with belt loop |
| Remarks: | Private-purchase hunting knife |

### TYPE 5

| | |
|---|---|
| Manufactured: | 1914 |
| Total Length: | 262 mm |
| Blade Length: | 127 mm |
| Hilt: | Staghorn with short curved quillons |
| Markings: | Anton Wingen, Solingen |
| Scabbard: | Blackened steel with belt loop |
| Remarks: | Private-purchase hunting knife |

### TYPE 6

| | |
|---|---|
| Manufactured: | 1914 |
| Total Length: | 275 mm |
| Blade Length: | 157 mm |
| Hilt: | Wood with S-shaped quillons |
| Markings: | None |
| Scabbard: | Brown leather with belt loop |
| Remarks: | Private-purchase hunting knife |

### TYPE 7

| | |
|---|---|
| Manufactured: | 1915 |
| Total Length: | 277 mm |
| Blade Length: | 155 mm |
| Hilt: | Wood grip in M.84/98 bayonet style |
| Markings: | Deutsche Maschinenfabrik A–G, Duisburg/DEMAG DRGM |
| Scabbard: | Blackened steel with belt loop |
| Remarks: | Bayonet/knife conversion |

## TYPE 8

| | |
|---|---|
| Manufactured: | 1915 |
| Total Length: | 256 mm |
| Blade Length: | 135 mm |
| Hilt: | Wood with 9 grooves and short quillon |
| Markings: | Feldzug 1915/ERN Solingen |
| Scabbard: | Blackened steel with belt loop |
| Remarks: | Sawtooth blade for cutting through barbed wire |

## TYPE 9

| | |
|---|---|
| Manufactured: | 1915 |
| Total Length: | 265 mm |
| Blade Length: | 149 mm |
| Hilt: | All steel with short straight quillons |
| Markings: | Ges. Gesch. and crowned 'F' inspection stamp |
| Scabbard: | Blackened steel with belt loop |
| Remarks: | None |

## TYPE 10

| | |
|---|---|
| Manufactured: | 1915 |
| Total Length: | 262 mm |
| Blade Length: | 152 mm |
| Hilt: | All steel, in a 'cranked' or 'bent' shape |
| Markings: | DEMAG, Duisburg DRGM/Ges. Gesch. |
| Scabbard: | Blackened steel with belt loop |
| Remarks: | Designed to fit as a bayonet on both the G.98 and K.98 rifles, hence the unique hilt shape |

## TYPE 11

| | |
|---|---|
| Manufactured: | 1915 |
| Total Length: | 262 mm |
| Blade Length: | 148 mm |
| Hilt: | Wood with 9 grooves and short quillon |
| Markings: | 2.K.103 unit mark and crowned 'C' inspection stamp |
| Scabbard: | Blackened steel with belt loop |
| Remarks: | None |

## TYPE 12

| | |
|---|---|
| Manufactured: | 1915 |
| Total Length: | 227 mm (extended) |
| Blade Length: | 100 mm |
| Hilt: | All steel |
| Markings: | Mercator Nahkampfer DRGM |
| Scabbard: | None |
| Remarks: | Locking knife with a retractable blade held in place by a clasp |

## TYPE 13

| | |
|---|---|
| Manufactured: | 1915 |
| Total Length: | 265 mm |
| Blade Length: | 141 mm |
| Hilt: | Artificial staghorn with S-shaped quillons |
| Markings: | Ed. Wüsthof, Solingen and trident trademark |
| Scabbard: | Brown leather with belt loop |
| Remarks: | None |

## TYPE 14

| | |
|---|---|
| Manufactured: | 1916 |
| Total Length: | 278 mm |
| Blade Length: | 146 mm |
| Hilt: | Wood with 9 grooves and short straight quillon |
| Markings: | Ernst Busch, Solingen |
| Scabbard: | Blackened steel with belt loop |
| Remarks: | None |

## TYPE 15

| | |
|---|---|
| Manufactured: | 1916 |
| Total Length: | 282 mm |
| Blade Length: | 154 mm |
| Hilt: | Wood with 9 grooves and cranked quillons |
| Markings: | DEMAG, Duisburg and crowned 'P' inspection stamp |
| Scabbard: | Blackened steel with belt loop |
| Remarks: | None |

## TYPE 16

| | |
|---|---|
| Manufactured: | 1916 |
| Total Length: | 271 mm |
| Blade Length: | 142 mm |
| Hilt: | Wood with 9 grooves and short bent quillon |
| Markings: | ERN Wald.Rhein, Rasiermesserfabrik |
| Scabbard: | Blackened steel with belt loop |
| Remarks: | None |

## TYPE 17

| | |
|---|---|
| Manufactured: | 1916 |
| Total Length: | 272 mm |
| Blade Length: | 143 mm |
| Hilt: | Wood with 9 grooves and short straight quillons |
| Markings: | ERN Wald.Rhein, Rasiermesserfabrik |
| Scabbard: | Blackened steel with belt loop |
| Remarks: | None |

## TYPE 18

| | |
|---|---|
| Manufactured: | 1916 |
| Total Length: | 265 mm |
| Blade Length: | 143 mm |
| Hilt: | Plain wood with short straight quillons |
| Markings: | 'Z' within crowned shield |
| Scabbard: | Blackened steel with belt loop |
| Remarks: | None |

## TYPE 19

| | |
|---|---|
| Manufactured: | 1916 |
| Total Length: | 277 mm |
| Blade Length: | 148 mm |
| Hilt: | Wood with 9 grooves and short straight quillons |
| Markings: | Gottlieb Hammesfahr, Solingen-Foche |
| Scabbard: | Blackened steel with belt loop |
| Remarks: | None |

## TYPE 20

| | |
|---|---|
| Manufactured: | 1916 |
| Total Length: | 270 mm |
| Blade Length: | 134 mm |
| Hilt: | Standard M.84/98 bayonet type with wood grips |
| Markings: | Rich.Abr.Herder, Solingen |
| Scabbard: | Standard M.84/98 bayonet type |
| Remarks: | Converted bayonet with shortened blade |

## TYPE 21

| | |
|---|---|
| Manufactured: | 1916 |
| Total Length: | 261 mm |
| Blade Length: | 143 mm |
| Hilt: | Banana-shaped wood with 11 grooves and S-shaped quillons |
| Markings: | Union, Zella St.B1. |
| Scabbard: | Blackened steel with belt loop |
| Remarks: | None |

## TYPE 22

| | |
|---|---|
| Manufactured: | 1916 |
| Total Length: | 275 mm |
| Blade Length: | 150 mm |
| Hilt: | Wood with 9 grooves and no cross-guard |
| Markings: | Feldzug 1916 |
| Scabbard: | Blackened steel with belt loop |
| Remarks: | Commercial knife adapted for combat use |

## TYPE 23

| | |
|---|---|
| Manufactured: | 1916 |
| Total Length: | 278 mm |
| Blade Length: | 150 mm |
| Hilt: | All-steel *ersatz* bayonet type with 9 grooves |
| Markings: | Ernst Busch, Solingen |
| Scabbard: | Blackened steel with belt loop |
| Remarks: | Converted bayonet with shortened blade |

## Type 24

| | |
|---|---|
| Manufactured: | 1917 |
| Total Length: | 260 mm |
| Blade Length: | 142 mm |
| Hilt: | All-steel *ersatz* bayonet type with 5 rivets |
| Markings: | Two storks and crowned 'F' inspection stamp |
| Scabbard: | Blackened steel with belt loop |
| Remarks: | Converted bayonet with knife blade fitted |

## Type 25

| | |
|---|---|
| Manufactured: | 1917 |
| Total Length: | 252 mm |
| Blade Length: | 157 mm |
| Hilt: | M.98 bayonet type with lanyard ring on the pommel |
| Markings: | Two squirrels over 'CE' |
| Scabbard: | Blackened steel with belt loop |
| Remarks: | Officer pattern |

## Type 26

| | |
|---|---|
| Manufactured: | 1918 |
| Total Length: | 268 mm |
| Blade Length: | 144 mm |
| Hilt: | Poorly finished wood with cheap zinc crossguard |
| Markings: | Crowned 'F' inspection stamp |
| Scabbard: | Blackened steel with belt loop |
| Remarks: | Inferior quality pattern |

## Type 27

| | |
|---|---|
| Manufactured: | 1919 |
| Total Length: | 178 mm |
| Blade Length: | 156 mm |
| Hilt: | M.98 bayonet type with chequered wooden grips |
| Markings: | Nahkampfer DRGM – Souvenir of Solingen |
| Scabbard: | Blackened steel with belt loop |
| Remarks: | War surplus item etched with 'souvenir' slogan in English and sold to Allied occupation troops and tourists |

# Fighting Knives, 1933–45

The Second World War German soldier had no less need of a fighting knife than his father, as the lightning victories of 1940 and 1941 gave way to a prolonged conflict of attrition in the east. Street battles in Stalingrad and trench assaults at the Kuban bridgehead saw hand-to-hand combat take on new and hellish proportions, with entire regiments engaged in close-quarter fighting for weeks on end. The *Nahkampfspange*, a new decoration featuring the traditional *Stosstrupp* emblems of grenade and bayonet, had to be created specifically to encourage and reward those involved. By 1944 it was being worn widely by corporals and colonels alike, such was the all-encompassing nature of man-to-man fighting in Russia.

The majority of *Wehrmacht* fighting knives were modelled on those of the 1914–18 period, the main difference being that Nazi versions tended to have smooth rather than grooved wooden grips. Moreover, the Second World War scabbard was normally fitted with a spring clip which facilitated attachment not only to the belt but also to any other convenient location such as the tunic front, equipment-supporting straps or jackboot.

A representative selection of Third Reich fighting knife patterns is described below.

## Type 1

| | |
|---|---|
| Manufactured: | 1935 |
| Total Length: | 282 mm |
| Blade Length: | 157 mm |
| Hilt: | M.98 bayonet type with chequered black plastic grips |
| Markings: | Original Eickhorn, Solingen and squirrel trademark |
| Scabbard: | Blackened steel with belt loop |
| Remarks: | Commercially produced for private purchase |

**161**. *The* Nahkampfspange, *or close-combat clasp, instituted on 25 November 1942 for award to those who had been engaged for long periods in hand-to-hand fighting at the front, took as its centrepiece the old* Stosstrupp *motifs of bayonet and grenade. This gold example was presented in recognition of over fifty days' close combat. Edged weapons of various types also featured on a number of other Nazi decorations, including the General Assault Badge, the Guerrilla Warfare Badge, the Wound Badge, the Demjansk Shield and a host of different awards bestowed 'with swords'.*

TYPE 2

| | |
|---|---|
| Manufactured: | 1935 |
| Total Length: | 275 mm |
| Blade Length: | 151 mm |
| Hilt: | Wood with 9 grooves and S-shaped quillons |
| Markings: | King's head and knight's helmet of WKC |
| Scabbard: | Blackened steel with belt loop |
| Remarks: | Identical in style to wooden-hilted 1914–18 patterns |

TYPE 3

| | |
|---|---|
| Manufactured: | 1935 |
| Total Length: | 263 mm |
| Blade Length: | 155 mm |
| Hilt: | Plain dark red plastic with three rivets and a short crossguard |
| Markings: | Puma, Solingen |
| Scabbard: | Blackened steel with spring clip |
| Remarks: | None |

**162**. *This standard* Wehrmacht *fighting knife, with the runic insignia from an SS 1933 dagger inserted into the handle, was captured by a US soldier serving in Italy during 1944.*

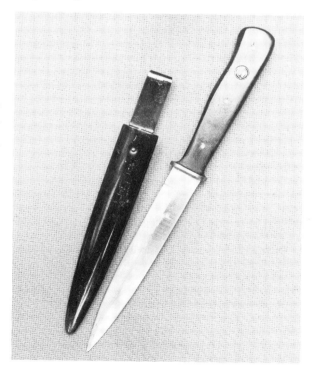

## TYPE 4

| | |
|---|---|
| Manufactured: | 1939 |
| Total Length: | 264 mm |
| Blade Length: | 147 mm |
| Hilt: | Plain wood with three rivets and short crossguard |
| Markings: | None |
| Scabbard: | Blackened steel with spring clip |
| Remarks: | None |

## TYPE 5

| | |
|---|---|
| Manufactured: | 1940 |
| Total Length: | 286 mm |
| Blade Length: | 172 mm |
| Hilt: | Polished wood with three rivets and a short crossguard |
| Markings: | Eagle over '5' *Luftwaffe* inspection stamp |
| Scabbard: | Blackened steel with spring clip |
| Remarks: | None |

## TYPE 6

| | |
|---|---|
| Manufactured: | 1940 |
| Total Length: | 285 mm |
| Blade Length: | 170 mm |
| Hilt: | Plain wood with three rivets and single quillon |
| Markings: | None |
| Scabbard: | Blackened steel with spring clip |
| Remarks: | None |

## TYPE 7

| | |
|---|---|
| Manufactured: | 1941 |
| Total Length: | 264 mm |
| Blade Length: | 150 mm |
| Hilt: | Plain wood with three rivets and no crossguard |
| Markings: | E. & F. Hörster, Solingen |
| Scabbard: | Blackened steel with spring clip |
| Remarks: | None |

## TYPE 8

| | |
|---|---|
| Manufactured: | 1942 |
| Total Length: | 253 mm |
| Blade Length: | 146 mm |
| Hilt: | M.98 bayonet type with black plastic grips and no quillon |
| Markings: | RZM M7/66 |
| Scabbard: | Blackened steel with belt loop |
| Remarks: | Converted surplus Hitler Youth *Fahrtenmesser*, with the grip insignia and quillon removed |

## TYPE 9

| | |
|---|---|
| Manufactured: | 1942 |
| Total Length: | 320 mm |
| Blade Length: | 200 mm |
| Hilt: | All steel, tubular in form, with a chequered grip and upswept crossguard |
| Markings: | Pet.Dan.Krebs, Solingen |
| Scabbard: | Wide black leather with steel strengthening plates and belt loop |
| Remarks: | Modelled on the British Fairbairn-Sykes fighting knife, examples of which were captured from Allied marine commandos during the ill-fated raid on Dieppe in August 1942 |

## TYPE 10

| | |
|---|---|
| Manufactured: | 1944 |
| Total Length: | 263 mm |
| Blade Length: | 155 mm |
| Hilt: | Plain dark red plastic containing numerous retractable tools, including a file, a tin-opener and a corkscrew |
| Markings: | Puma, Solingen |
| Scabbard: | Blackened steel with long boot clip extending above the scabbard mouth |
| Remarks: | Updated variant of Type 3 knife |

## TYPE 11

| | |
|---|---|
| Manufactured: | 1945 |
| Total Length: | 250 mm |
| Blade Length: | 130 mm |
| Hilt: | Crudely fashioned wood with two rivets and no crossguard |
| Markings: | None |
| Scabbard: | Blackened steel with belt loop |
| Remarks: | Inferior-quality conversion from 1914–18 stock parts |

Due to the ready supply of surviving dress daggers at the end of the Second World War, there was no demand from Allied occupation troops for 'souvenir' Nazi fighting knives. Most were simply taken into civilian use or destroyed.

# UTILITY KNIVES

Utility knives were designed as tools rather than weapons but needless to say the exigencies of warfare resulted in many of them being used in hand-to-hand combat against the enemy. The examples which follow give a good idea of the variety of utility knife patterns produced during the Third Reich.

### THE *WAFFEN-LOESCHE* KNIFE

This sidearm comprised a 34 cm knife with a simple staghorn grip, and was held in a bayonet-type blackened steel scabbard suspended from a brown leather frog. It derived its unofficial name from the fact that all examples bore the distributor's mark 'Waffen-Loesche, Berlin'. The knife was issued on a very restricted basis to men attached to seaplane squadrons, shipboard reconnaissance units and certain other specialist flying formations of the *Luftwaffe*. It was superseded by the ubiquitous flight utility knife in 1939.

### THE FLIGHT UTILITY KNIFE

Just prior to the outbreak of the Second World War, the Solingen firm of Paul Weyersberg developed a safety knife to be carried by all aircrew for emergency use when on missions. It was constructed so that the 107 mm blade was totally enclosed within the wooden hilt when not in use. The blade's extension and retraction was achieved solely by the force of gravity, being effected by pressing a spring clip above the hilt whilst pointing the knife downwards or upwards as appropriate. This allowed the knife to be opened or closed in a one-handed operation. A marlin spike was fitted below the blade housing, and the earliest examples could be taken apart for ease of cleaning and repair.

The flight utility knife was soon being produced by several different firms and was distributed widely, not only to airmen but also to all German paratroopers. This gave rise to its being universally referred to as the *Fallschirmjäger-Schwerkraftmesser*, or 'paratrooper gravity knife'. The functional design proved so successful that in 1941 the British unashamedly copied it in the production of a similar weapon for their own airborne forces.

### THE *LUFTWAFFE* SURVIVAL MACHETE

A massive machete was issued as standard equipment to *Luftwaffe* bomber crews. It formed part of each aircraft's inventory, and was to be employed as a utility tool or survival weapon in the event of the crew being grounded in hostile conditions. Produced by Alexander Coppel, the machete had a large bolo-shaped blade, a grooved wooden grip

*163. Flight utility knives, by SMF. The example on the left has nickel-plated fittings and cannot be disassembled. That on the right has blued components and can be taken apart for cleaning, the stamped arrow pointing to the square release button.*

and S-shaped quillons, and was contained in a blued steel scabbard. Although not intended to be worn from the waist, it was supplied with a green canvas webbing frog.

There were several recorded instances of the machete being used to extremely good effect by downed crews engaged in close-quarter fighting on the eastern front.

### The Diver's Knife

A special sidearm was produced for daily working use by naval divers and frogmen. Manufactured by the Henckels and Koeping firms, it featured a heavy cylindrical ribbed hilt in solid brass and an 18 cm stainless steel blade with a partially serrated edge. The knife was screwed into its tubular brass scabbard, and had a rubber seal at the ricasso to keep the blade area watertight.

### The Combination Knife

As early as 1935, Gebrüder Gräfrath of Solingen made an experimental bayonet containing tools in the hilt, and the Puma firm manufactured a fighting knife with similar accessories at the beginning of

1944. These ideas were developed by Eickhorn during mid-1944 to produce a single combination weapon designed to function equally as:

- a bayonet for the Kar. 98k rifle
- a fighting knife
- a utility knife

The 29 cm sidearm had brown bakelite grips and was held in a blued steel scabbard with spring clip attachment. A removable multi-purpose tool component contained within the hilt comprised a penknife blade, a bottle-opener, a tin-opener, a corkscrew, a screwdriver and a marlin spike.

The M.44 combination knife represented the ultimate in practicality, but did not see widespread distribution before the end of the war.

German fighting and utility knives therefore came in many shapes and sizes. As well as regulation types, an array of hunting daggers, cut-down bayonets and field-made items were also employed. Even surplus dress weapons were eventually pressed into use for practical purposes at the front line, with at least one *Waffen-SS* NCO known to have worn an old 1933-pattern *SS* dagger attached to his assault pack while on active service in France in 1944. The fortunes of the Nazi edged weapon had, indeed, turned full circle.

**164**. *This diver's knife is accompanied by an exquisitely fashioned miniature.*

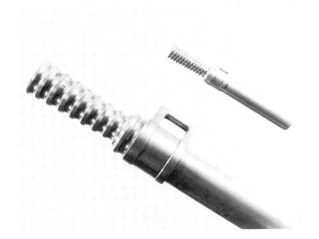

# PART FIVE

# MAKERS' AND OTHER MARKS

## TRADEMARKS

Well over 100 different firms, ranging from one-man 'cottage' enterprises to large industrial concerns, were engaged in the manufacture of daggers, swords, bayonets and knives during the Third Reich. The vast majority of these companies were located in Solingen, but others could be found in Berlin, Brotterode, Göttingen, Magdeburg, Marienthal, Neustadt, Riemberg, Schmalkalden, St Christophen, Steinbach, Steyr, Suhl, Tuttlingen and Vienna.

A self-explanatory maker's trademark, usually comprising a name, logo and town designation, was commonly etched or stamped on the reverse ricasso of each edged weapon's blade. Pieces produced during the war years were occasionally found without such marks, but these were the exception rather than the rule.

## RZM CODES

During the 1935–6 period the *RZM*, or *Reichszeugmeisterei*, a body which had been set up as early as 1 April 1929 to supervise the production and pricing of all Nazi party uniform items, was effectively reorganised by Richard Büchner and began to extend its influence into the realm of edged weapons. The basic functions of the *RZM* were to see that *NSDAP* contracts went to Aryan firms and to ensure that the final products were of a high standard, yet priced to suit the pocket of the average Party member. It also acted as a clearing house between manufacturers on the one hand and wholesalers and retailers on the other. On 16 March 1935, contract numbers were introduced and awarded to every *RZM*-approved company. After that date, *RZM* numbers were supposed to replace makers' trademarks on all *NSDAP* accoutrements. However, this directive took some time to work its way through the system and many firms continued to use their company logos on Nazi Party items well into 1936.

By the following year, the list of firms approved by the *RZM* had grown so considerably that it had to be divided into the following groups, depending upon function:

A – *Ausrüstung*: equipment maker
B – *Baumwolle*: cotton fabric manufacturer
D – *Dienstkleidung*: service clothing producer
G – *Grosshandel*: wholesaler
H – *Handelsvertreter*: maker's agent
K – *Kleidereinzelhandel*: clothing retailer
L – *Leder*: leather goods maker
M – *Metall*: metalware manufacturer
W – *Wolle*: wool fabric producer

| | | | | | |
|---|---|---|---|---|---|
| | **Karl Böcker**<br>Solingen, Schützenstr. 24  M 7/45 | | **Gebr. Born**<br>Solingen  M 7/82 | CHRISTIANS SOLINGEN | **GEBR. CHRISTIANS CHRISTIANSWERK**<br>Solingen  M 7/1 |
| | **Carl Eichhorn**<br>Solingen  M 7/66 | A E S | **Arthur Evertz**<br>Solingen  M 7/85 | Granrise | **Gebr. Gräfrath**<br>Solingen·Widdert  M 7/30 |
| | **Josef Hack** Ges. m. b. H.<br>Steyr (Ober·Donau)  M 7/103 | HH | **Herm. Hahn**<br>Solingen·Wald  M 7/71 | | **Carl Halbach**<br>Solingen  M 7/23 |
| NEROSTA | **Gottlieb Hammesfahr**<br>Solingen·Foche  M 7/67 | | **Hartkopf & Co.**<br>Solingen  M 7/40 | M &H | **Herberts & Meurer**<br>Solingen·Gräfrath  M 7/52 |
| | **Friedr. Herder A. S.**<br>Constantwerk<br>Solingen  M 7/49 | | **H. HERDER**<br>Solingen  M 7/111 | | **Richard Abr. Herder**<br>Solingen  M 7/18 |
| | **Robert Herder**<br>Solingen·Ohligs  M 7/55 | | **F. W. Höller**<br>Solingen  M 7/33 | HS | **E. & F. Hörster**<br>Solingen  M 7/36 |
| | **Juliuswerk**<br>**J. Schmidt & Söhne**<br>Riemberg i. Schlesien  M 7/88 | KARL KALDENBACH | **Karl Rob. Kaldenbach**<br>Solingen·Gräfrath  M 7/72 | | **Robert Klaas**<br>Solingen·Ohligs  M 7/37 |
| | **Klittermann & Moog**<br>G. m. b. H.<br>Haan bei Solingen  M 7/29 | SOLINGEN | **E. Knecht & Co.**<br>Solingen  M 7/11 | | **F. Koeller & Co.**<br>Solingen·Ohligs  M 7/97 |
| GUSTAV L. KÖLLER SOLINGEN-WALD | **Gustav L. Köller**<br>**Nachf.**<br>Solingen·Wald  M 7/60 | | **Herm. Konejung A.·G.**<br>Solingen  M 7/7 | | **Carl Julius Krebs**<br>Solingen  M 7/5 |
| | **Pet. Dan. Krebs**<br>Solingen  M 7/92 | | **Lauterjung & Co.**<br>Tiger·Stahlwaren u.·Waffenfabrik<br>Solingen  M 7/68 | C & R LINDER | **Carl & Robert Linder**<br>Solingen·Weyer  M 7/15 |
| Senta | **Herm. Linder Söhne**<br>Solingen  M 7/78 | LINDR | **Hugo Linder**<br>**C. W. Sohn**<br>Solingen·Weyer  M 7/114 | | **P. D. Lüneschloß**<br>Solingen  M 7/14 |
| M | **August Merten Ww.**<br>Solingen·Gräfrath  M 7/21 | | **Friedr. Plücker jr.**<br>Solingen·Gräfrath  M 7/62 | PUMA | **Puma-Werk**<br>Lauterjung & Sohn<br>Solingen  M 7/27 |
| REMEVE | **Cuno Remscheid & Co.**<br>Solingen·Aufderhöhe  M 7/86 | | **Kuno Ritter**<br>Solingen·Gräfrath  M 7/3 | SMF | **Solinger Metallwaren-**<br>**Fabrik Stöcker & Co.**<br>Solingen  M 7/9 |
| | **C. Gustav Spitzer K.·G.**<br>Solingen  M 7/80 | | **C. D. Schaaff**<br>Solingen  M 7/56 | | **Carl Schmidt Sohn**<br>**K.·G.**<br>Solingen  M 7/84 |
| | **J. A. Schmidt & Söhne**<br>Solingen  M 7/95 | | **Rudolf Schmidt**<br>Solingen  M 7/41 | | **Artur Schüttelhöfer**<br>**& Co.**<br>Solingen·Wald  M 7/13 |
| Engelswerk | **Karl Tiegel**<br>Riemberg Bez. Breslau  M 7/81 | | **Emil Voos**<br>Solingen  M 7/2 | WMW | **WAFFENFABRIK**<br>**MAX WEYERSBERG**<br>Solingen  M 7/12 |
| W | **Wilh. Wagner**<br>Solingen·Merscheid  M 7/25 | | **Paul Weyersberg & Co.**<br>Solingen  M 7/43 | KRONECK | **Ernst Erich Witte**<br>Solingen  M 7/98 |
| WKC | **WKC·Waffenfabrik**<br>(Gesellschaft m. beschränkt. Haftung<br>Solingen·Wald  M 7/42 | SPRINGER | **Carl Wüsthof**<br>**Gladiatorwerk**<br>Solingen  M 7/112 | | **Ed. Wüsthof**<br>**Dreizackwerk**<br>Solingen  M 7/19 |

**165**. *Some of the wide range of eye-catching trademarks employed by the Third Reich's edged weapons makers. Artistic monograms, ingenious plays on words (such as the 'crab' logo of Peter Krebs), and snappy abbreviations (like 'SMF' for 'Solinger Metallwarenfabrik Stöcker & Co.', or 'Alcoso' for 'Alexander Coppel, Solingen') were all used to good effect in the design of such marks.*

**166**. *This form of the Eickhorn trademark was current between 1933 and 1938. The SA chained Honour Dagger was consequently one of the last items on which it was used.*

**167**. *The trademark of Paul Weyersberg etched into the blade of an RLB 1938 knife.*

**168**. *The E. & F. Hörster trademark on the head of a Fire Brigade presentation axe.*

Each group was in turn subdivided into product categories. Those for metalware manufacturers were:

M1 – insignia (except day badges)

M2 – subcontractors

M3 – emblems (standard tops, etc.)

M4 – belt buckles

M5 – uniform fittings

M5a – buttons

M5b – side hooks

M5c – snap hooks

M5d – other metal fittings

M6 – aluminium goods

M7 – daggers

M8 – metal accessories (tent pegs, etc.)

M9 – day badges

M10 – musical instruments

New contract code numbers were then issued, prefixed by the relevant category designation. So far as makers of *NSDAP* daggers were concerned, the designation was always 'M7'. Thus the Eickhorn company, which was allocated the *RZM* contract number 66, was obliged to replace its well-known squirrel logo with the code '*RZM* M7/66' on its Nazi Party daggers.

*RZM* codes were not used on swords or bayonets, or on the edged weapons of the *Wehrmacht*

or government authorities. Neither did they feature on the daggers of the *NPEA*, which was considered to be an organ of the state education system rather than of the party, or on those of the public-service *RAD* or the *Luftwaffe*-related *NSFK*. In effect, *RZM* codes were restricted to the daggers of the *SA*, *SS*, *NSKK* and *HJ*. Honour Daggers of these organisations did not fall within the remit of the *RZM* and so continued to bear makers' trademarks.

The following table lists the *RZM* 'M7' code numbers known to have been issued, and the dagger-making firms to which they referred. All these companies were based in Solingen, unless otherwise indicated. The *RZM* logo, as shown, always preceded the code number. (RZM)

**169**. *An unissued* SA *1933 dagger by Richard Abraham Herder, complete with paper packing bag. As well as the Herder trademark, the blade bears the firm's* RZM *designation 'M7/18', and date '1939'. The use of a company logo in conjunction with an* RZM *mark at this late date is most unusual.*

| *RZM* CODE | FIRM |
|---|---|
| M7/1 | Gebrüder Christians, Christianswerk |
| 2 | Emil Voos |
| 3 | Kuno Ritter |
| 4 | August Müller |
| 5 | Carl Julius Krebs |
| 6 | H. & F. Lauterjung |
| 7 | Hermann Konejung |
| 8 | Eduard Gembruch |
| 9 | Solinger Metallwarenfabrik Stöcker & Co. (SMF) |
| 10 | J. A. Henckels, Zwillingswerk |
| 11 | E. Knecht & Co. |
| 12 | Waffenfabrik Max Weyersberg (WMW) |
| 13 | Artur Schüttelhöfer & Co. |
| 14 | P.D. Lüneschloss |
| 15 | Carl und Robert Linder |
| 16 | Justus Berger & Co., Justinuswerk |
| 17 | A. Werth |
| 18 | Richard Abraham Herder |
| 19 | Eduard Wüsthof, Dreizackwerk |
| 20 | Ernst Mandewirth |
| 21 | Hermann Schneider |
| 22 | Wilhelm Weltersbach, Laufbrunnenwerk |
| 23 | Carl Halbach |
| 24 | Reinhard Weyersberg |
| 25 | Wilhelm Wagner |
| 26 | Jacobs & Co. |

| | | | |
|---|---|---|---|
| 27 | Lauterjung & Sohn, Pumawerk | 73 | F. & A. Helbig, Steinbach |
| 28 | Gustav Felix, Gloriawerk | 74 | Friedrich August Schmitz |
| 29 | Klittermann & Moog | 75 | Böker & Co. |
| 30 | Gebrüder Gräfrath, Gränisowerk | 76 | Herbeck & Meyer |
| 31 | August Merten | 77 | Gustav Schneider |
| 32 | Robert Müller & Sohn | 78 | Hermann Linder & Söhne, Sentawerk |
| 33 | F.W. Höller | 79 | C. Bertram Reinhardt |
| 34 | C. Rudolf Jacobs | 80 | C. Gustav Spitzer |
| 35 | Wilhelm Halback | 81 | Karl Tiegel, Reimberg |
| 36 | E. & F. Hörster | 82 | Gebrüder Born |
| 37 | Robert Klaas | 83 | Richard Plümacher |
| 38 | Paul Seilheimer | 84 | Karl Schmidt & Sohn |
| 39 | Franz Steinhoff | 85 | Arthur Evertz |
| 40 | Hartkopf & Co. | 86 | Cuno Remscheid & Co., Remevewerk |
| 41 | Rudolf Schmidt | 87 | Malsch & Ambronn |
| 42 | Weyersberg, Kirschbaum & Co. (WKC) | 88 | J. Schmidt & Söhne, Juliuswerk, Riemberg |
| 43 | Paul Weyersberg & Co. | 89 | Ernst Mandewirth |
| 44 | F.W. Backhaus & Co. | 90 | Eickelnberg & Mack |
| 45 | Karl Böcker | 91 | Carl Malsch-Spitzer, Steinbach |
| 46 | Emil Gierling | 92 | Peter Daniel Krebs |
| 47 | Paul Ebel | 93 | Ewald Cleff |
| 48 | Otto Simon | 94 | Gebrüder Bell |
| 49 | Friedrich Herder | 95 | J.A. Schmidt & Söhne |
| 50 | Gebrüder Heller, Marienthal | 96 | Drees & Sohn |
| 51 | Anton Wingen Jr, Othellowerk | 97 | F. Koeller & Co. |
| 52 | Herbertz & Maurer | 98 | Ernst Erich Witte, Kroneckwerk |
| 53 | Gustav Weyersberg | 99 | Franz Weinrank, Vienna |
| 54 | Gottfried Müller | 100 | Franz Pils & Söhne, Steyr |
| 55 | Robert Herder | 101 | Fritz Weber, Vienna |
| 56 | C.D. Schaaff | 102 | Peter Prass |
| 57 | Peter Lungstrass | 103 | Josef Hack, Steyr |
| 58 | Louis Perlmann | 104 | Ludwig Zeitler, Vienna |
| 59 | C. Lütters & Co. | 105 | Rudolf Wurzer, St Christophen |
| 60 | Gustav L. Köller | 106 | Georg Kerschbaumer, Steinbach |
| 61 | Carl Tillmann & Söhne | 107 | Bremshey & Co. |
| 62 | Friedrich Plücker Jr | 108 | Josef Wolf |
| 63 | Herder & Engels | 109 | Fritz Balke |
| 64 | Friedrich Geigis | 110 | Gebrüder Knoth |
| 65 | Carl Heidelberg | 111 | H. Herder |
| 66 | Carl Eickhorn | 112 | Carl Wüsthof, Gladiatorwerk |
| 67 | Gottlieb Hammesfahr | 113 | Wilhelm Otto |
| 68 | Lauterjung & Co., Tigerwerk | 114 | Hugo Linder C.W. Sohn, Linorwerk |
| 69 | H.A. Erbe, Schmalkalden | 115 | Wilhelm Goeddertz |
| 70 | David Malsch, Steinbach | 116 | Franz Frenzel |
| 71 | Hermann Hahn | 117 | Rudolf & Karl Kraus |
| 72 | Karl Robert Kaldenbach | 118 | Jacobs & Co. |

**170**. *The* Waffenamt *stamp on this Navy sword hanger shows that the piece passed inspection in Hamburg in 1936.*

## SS-RZM CODES

In 1935, the *SS* devised its own peculiar system of code designations for the relatively small number of *RZM*-approved companies which were at that time contracted by Himmler to produce regulation *SS* uniform items. Such codes took the form of the *SS* runes, a maker's number, a year suffix and the *RZM* logo. So far as dagger manufacturers were concerned, the following *SS-RZM* numbers are known to have been allocated:

| | |
|---|---|
| 324 | Jacobs & Co. |
| 941 | Carl Eickhorn |
| 1051 | Robert Klaas |
| 1053 | P. D. Lüneschloss |
| 1164 | Böker & Co. |
| 1166 | Emil Voos |
| 1197 | C. Gustav Spitzer |
| 1211 | Ernst Pack & Söhne |
| 1221 | Carl Malsch-Spitzer |

Thus the code (SS) 941/36 (RZM) would relate to an *SS* dagger made by Eickhorn in 1936.

As contract documents were often signed for five or more years, these firms' *SS-RZM* numbers continued to be used until their date of expiry, with only the code's year suffix changing on an annual basis. Manufacturers newly contracted to produce *SS* daggers after full *RZM* standardisation had been achieved in 1937 were allocated regular 'M7' codes, not the *SS* type.

## DISTRIBUTORS' MARKS

A small minority of edged weapons were marked with details of their retail distributor, usually in the form of a name and town location stamped on to the back edge of the blade. In a very few cases the distributor's logo was more akin to that of a manufacturer, an example being the mark of Karl Burgsmüller which was etched on the reverse ricasso of many *NPEA* daggers.

## INSPECTION AND PROOF MARKS

*WAFFENAMT STAMPS*

*Wehrmacht*-issue weapons had to pass rigorous examination by military armourers, whose tiny inspection, or *Waffenamt*, stamps were thereafter applied to one or more of a number of locations, including the blade and hilt. The *Waffenamt* stamps on Army and Navy pieces issued prior to May 1934 featured the Weimar Republic-style eagle without a swastika, while later marks took the Nazi form. Naval items were distinguished by the usual 'M' (*Marine*) designation below the eagle, while the *Luftwaffe* employed its own distinctive pattern of *Waffenamt* stamp. All of these could be with or without armoury serial numbers. Examples of each style are shown below.

| STAMP | BRANCH OF SERVICE |
|---|---|
| WaA138 | Army, 1933–4 |
| M | Navy, 1933–4 |
| WaA138 | Army, 1934–45 |
| M | Navy, 1934–45 |
| 5 | *Luftwaffe*, 1935–45 |

## ORGANISATIONAL MARKS

The headquarters abbreviation inspection stamps used by the *SA*, *SS* and *NSKK* between 1933 and 1936 have already been covered in the appropriate parts of the text. Other organisational inspection logos which may be encountered are shown below, together with explanations:

 *SS* inspection stamp. This mark was introduced by Himmler around 1936 to indicate that a design had been commissioned and approved by the *SS*, according to its own strict artistic standards. It featured on porcelain, jewellery and similar items produced by or for the *SS*. The logo was also used as an inspection stamp on *SS* chained daggers, *SS* swords and Police swords, which did not fall within the remit of the *RZM*.

 *RAD* approval mark. The initials '*RJAD*' stood for *Reichsjugendarbeitsdienst*, a term used for only a short period between 1933 and 1934.

 *DLV* inspection stamp

 *NSFK* inspection stamp

It is worth noting that inspection and proof marks normally measured only a few millimetres in diameter, and were frequently mis-struck when applied. Consequently, they are best examined under strong magnification.

# PROPERTY AND ACCOUNTABILITY MARKS

Edged weapons which were issued from unit stores to personnel on a temporary basis, for use and return as required, were engraved or stamped with organisational property marks and/or accountability serial numbers. Examples included:

**171.** *The suspension clip on the vertical hanger of an SS 1933 dagger, clearly showing the 'DRGM' mark.*

- *Luftwaffe*, *DLV*, *NPEA*, *Postschutz* and *Teno* daggers
- Army and *Luftwaffe* swords
- *Wehrmacht* and Police bayonets

These have been described under the relevant section headings.

## OTHER MARKS

The following miscellaneous marks were also applied from time to time on patented items, those made under licence, and certain high-grade presentation pieces:

| MARK | EXPLANATION |
|---|---|
| *DRGM* | *Deutsches Reichsgebrauchsmuster* (state registered design) |
| *DRP* | *Deutsches Reichspatent* (patented). This mark usually appeared on Army lion-head sword patterns which were the exclusive product of a single manufacturer. It should not be confused with the '*DRP*' property mark of the *Deutsche Reichspost*, which was stamped on *Postschutz* daggers and bayonets. |
| *GES.GESCH.* | *Gesetzlich geschützt* (patent pending) |
| *SILBER 935* | 93.5 silver content. Other silver contents were 800, 835, 900 and 990. |

# PART SIX

# SOLINGEN

# SALESMANSHIP

The selling of Third Reich edged weapons and their many accessories was big business for all the contemporary commercial enterprises involved, whether they were makers, wholesalers or retailers. To advertise their wares, the firms concerned generated considerable quantities of sales material, including lavishly illustrated catalogues and brochures, each identifying dozens of different patterns of dagger, sword and bayonet by model number, patent name where applicable, authorised wearer and options available at extra cost. There were also attractive charts, calendars and window displays, backed up by competitive price lists aimed at persuading prospective buyers to make the 'right' choice. The Alcoso, Höller, Hörster, Krebs, Pack and WKC companies devised particularly impressive promotional publications, but none could surpass those of Carl Eickhorn, widely recognised as the premier cutler of Hitler's Germany.

The Eickhorn company was founded in 1865 and so celebrated its seventy-fifth anniversary in 1940. To mark the event, a commemorative book entitled *Leisten und Dienen* ('Providing and Serving') was published, tracing the history of the firm and giving an important insight into the dagger- and sword-manufacturing business during the Nazi regime. It relates how the year 1933 heralded the dawning of a new era for German edged weapon producers, with a degree of prosperity

*172. A WKC sales catalogue, showing the company's distinctive 'knight's helm' logo. With the death of Hermann Weyersberg in 1883, his firm amalgamated with that of W.R. Kirschbaum to form Weyersberg, Kirschbaum & Co., which became one of Solingen's principal edged weapon manufacturers.*

*173. A small selection of the vast array of publicity material produced by Solingen's armourers.*

business and associated matters. For example, they listed the four main wholesale distributors of private-purchase daggers and swords as:

- Albert Kreth, Bachstrasse 68, Hamburg
- Walter Kullak, Hasenheide 71, Berlin
- Kilian Spatteneder, Steinheilstrasse 15/1, Munich
- Viktor Vitali, Hasnerstrasse 54, Vienna

Prices, the latest patent developments, recommended sword lengths for buyers of different heights and even postage costs were all covered in these publications.

Every manufacturer took great care to protect his finished products when they were being shipped to wholesalers or retailers. The extent of packing varied from firm to firm, but scabbards were normally wrapped in strong brown paper and then sealed in stiff cardboard cylinders. Full-length cloth covers for everyday use were either provided with the sidearm or made available as private-purchase options, and companies did not hesitate to display their trademarks prominently on such accessories. Most daggers, swords and bayonets came complete with guarantee labels or tags, the Eickhorn version taking the form of a red metal 'seal' which contained a small certificate giving details of the unequivocal quality guarantee extended by the firm.

Solingen's blade-makers employed teams of travelling salesmen who toured uniform shops, military barracks, officer training schools, etc., soliciting individual or group orders for their companies' products. Rather than carrying unwieldy regulation sidearms with them on such trips, some of these individuals were equipped with a very restricted selection of highly detailed miniature edged weapons which they used as samples and as gifts to be presented to particularly valued customers. Miniatures could range in scale from about a third to three-quarters of the dimensions of full-sized production pieces and were often employed as paper knives, or mounted on small wooden plinths in the shape of anchors or propellors to be displayed as desk ornaments or paperweights. Most could be completely disassembled into the same

throughout all branches of the industry which had never been experienced before. Profits rocketed, Solingen's factories expanded rapidly, and the working conditions of employees were improved out of all recognition with the installation of modern canteens, comfortable furnishings and proper sanitation. Consequently, in a pattern repeated across the Reich, the formerly downtrodden workers of the Ruhr, who had been ardent Communists during the 1920s, were 'converted' *en masse* to become staunch supporters of the Nazi Party. Hitler's programme of reducing unemployment through rearmament and public works had paid real dividends, restoring national pride and securing the long-term prospects of the *NSDAP*.

Eickhorn's more general catalogues, and those of its competitors, were compiled not only as sales material but also as ready references on the cutlery

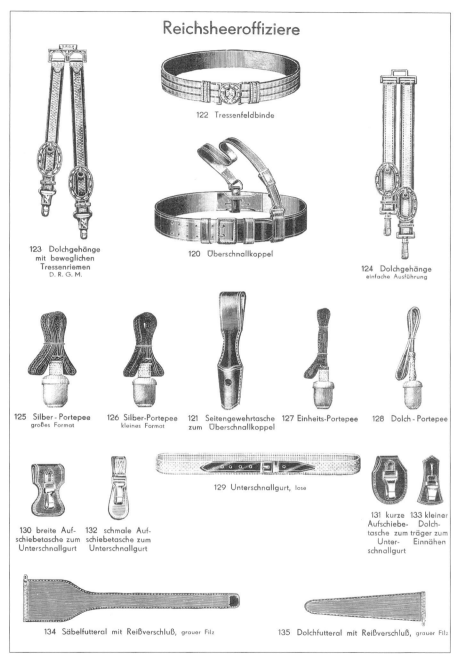

## Reichsheeroffiziere

122 Tressenfeldbinde

123 Dolchgehänge
mit beweglichen
Tressenriemen
D. R. G. M.

120 Überschnallkoppel

124 Dolchgehänge
einfache Ausführung

125 Silber - Portepee
großes Format

126 Silber-Portepee
kleines Format

121 Seitengewehrtasche
zum Überschnallkoppel

127 Einheits-Portepee

128 Dolch - Portepee

129 Unterschnallgurt, lose

130 breite Auf-
schiebetasche zum
Unterschnallgurt

132 schmale Auf-
schiebetasche zum
Unterschnallgurt

131 kurze
Aufschiebe-
tasche zum
Unter-
schnallgurt

133 kleiner
Dolch-
träger zum
Einnähen

134 Säbelfutteral mit Reißverschluß, grauer Filz

135 Dolchfutteral mit Reißverschluß, grauer Filz

**174**. *The Solingen firm of Röder & Co. sold a wide selection of edged weapons accoutrements, including this range aimed at Army officers. Of particular note are the grey felt zip-fastening storage bags for sabres and daggers illustrated at the bottom of the page. Grey-coloured bags were clearly not the preserve of the SS, as is often maintained.*

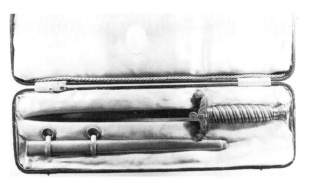

**175**. *Before 1939, Eickhorn delivered even the most basic of its* Wehrmacht *daggers in elaborate cases, most of which were simply discarded by their buyers.*

**176**. *A packing bag for an* SA *dagger, bearing the style of Eickhorn logo used from 1941.*

**177**. *Army dress bayonets were usually shipped in strong brown paper bags. The printed designation* Seitengewehr Ganze Länge 40 cm *denotes a sidearm for enlisted ranks.*

**178**. *An Eickhorn quality guarantee tag.*

components as their regulation counterparts and some featured suitably etched blades. A few were even equipped with scaled-down portepée knots and hanging straps, and were housed in little presentation cases specially recessed to accommodate them.

Reduced versions of the following items are known to have been produced by the makers indicated:

- Army 1935 dagger (Alcoso)
- Navy 1938 dagger (Alcoso)
- *Luftwaffe* 1935 dagger (Alcoso)

**179**. *A miniature Army 1935 dagger, shown alongside its full-size counterpart for comparison.*

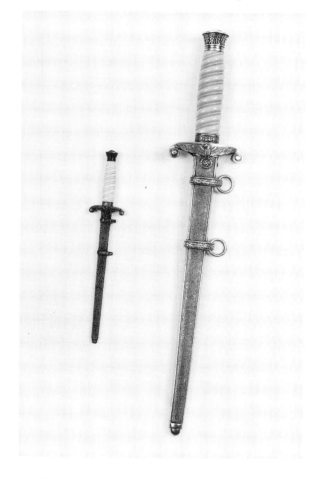

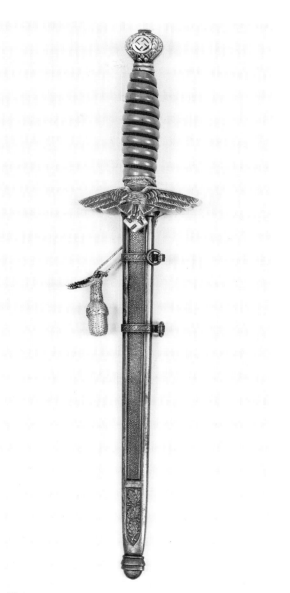

**180**. *This miniature* Luftwaffe *1937 dagger is complete with a tiny portepée knot, tied in the correct way.*

- *Luftwaffe* 1937 dagger (E. & F. Hörster; SMF)
- *SS* 1936 chained dagger (unmarked: probably Eickhorn)
- *RAD* leader's 1937 hewer (J.A. Henckels)
- Hunting Association 1936 cutlass (Alcoso)

- *Luftwaffe* 1934 sword (E. & F. Hörster)
- *SS* officer's 1936 sword (unmarked: probably Peter Daniel Krebs)
- *Wehrmacht* dress bayonet (SMF; WKC)
- Diver's knife (Henckels and Koeping)

The majority of armed forces miniatures were unmarked, so a number of other companies were probably involved in their manufacture. *SS, RAD* and Hunting Association examples were extremely rare and may, in fact, have been intended solely as gifts rather than sales promotions.

It is clear that miniatures were aimed primarily at the military officer class rather than at those belonging to political and civil organisations.

In addition to catalogues, professional journals such as the magazine *Die Klinge* and the Solingen Chamber of Commerce periodical *Blanke Waffen* circulated widely, serving to encourage competition and spread best practice amongst cutlers. Surviving examples of these publications and associated sales material, especially illustrated types, have proved to be invaluable sources of information, facilitating the accurate identification of unusual Nazi edged weapons. For example, the so-called Armistice Sabre, with its langet bearing an eagle in flight clutching an olive branch of peace rather than a swastika, which was previously believed to have been produced to mark the end of hostilities in 1918 and again in 1945, was recently revealed from a 1937 Paul Weyersberg catalogue as simply another variant Third Reich sword model made available to army officers for private purchase. In a similar vein, *Die Klinge* and *Blanke Waffen* have confirmed the production in 1940 and 1941 of the Diplomatic Corps bayonet and various prototype Army *Säbel* designs, while the uniform producers' journal *Uniformen-Markt* is the sole source proving the existence of the Railway Service dagger in a finished form in May, 1941.

Even local daily newspapers such as the *Solinger Tageblatt* carried regular features on a variety of subjects related to edged weapons, from announcing the formation and composition of an *SA* Dagger Production Committee on 12 February 1934 to illustrating the correct manner of tying a portepée knot to the *SS* chained dagger in February, 1943.

In view of their undoubted worth as reference material, the ephemera and publicity paraphernalia used by Solingen salesmen have become almost as valued as the merchandise they were originally intended to promote.

# Appendix I:

# Chronology

This table charts the creation and development of German edged weapons between 30 January 1933, when Hitler came to power, and 23 May 1945 when the Third Reich officially ended. For ease of reference, selected significant historical events are included and shown in parenthesis.

It is clear that the fortunes of the cutlery industry mirrored those of the Nazi regime itself. The vast majority of dress daggers and swords were designed and manufactured during the first six years of Hitler's rule, a peaceful era of growing confidence and prosperity, while the bleaker Second World War period witnessed a sharp decline in the use of ceremonial pieces and a concentration on the production of more practical combat bayonets and fighting knives. The latter continued to be made in large quantities until the virtual destruction of Solingen by Allied bombing in August and September 1944.

## – 1933 –

| | |
|---|---|
| 30 Jan. | (HITLER BECOMES CHANCELLOR) |
| 15 Feb. | Hitler Youth Knife introduced |
| 14 Jul. | (*NSDAP* BECOMES THE ONLY LEGAL POLITICAL PARTY IN GERMANY) |
| 10 Aug. | Design of miner's sabre standardised |
| 10 Aug. | Length of Police bayonet standardised |
| 7 Oct. | *SA-Feldjägerkorps* sabre introduced |
| 7 Oct. | *SA-Feldjägerkorps* bayonet introduced |
| 7 Dec. | *SA-Stabswache Göring* sabre introduced |
| 15 Dec. | *SA* dagger introduced |
| 15 Dec. | *SS* dagger introduced |

## – 1934 –

| | |
|---|---|
| 3 Feb. | *SA* Röhm Honour Dagger introduced |
| 3 Feb. | *SS* Röhm Honour Dagger introduced |
| 12 Feb. | Solingen's factory owners form an *SA* Dagger Production Committee to co-ordinate the mass manufacture of *SA* and *SS* daggers |
| 17 Feb. | Private purchase or 'trading in' of *SS* daggers on the open market forbidden |
| 21 Feb. | *Stabschef* Röhm authorises his senior *SA* and *SS* commanders to commission production of their own Honour Daggers for presentation to deserving personnel on a local basis |
| 17 Mar. | *RAD* hewer introduced |
| 30 Mar. | *DLV* dagger introduced |
| 30 Mar. | *DLV* knife introduced |
| 30 Mar. | *DLV* sword introduced |
| 23 Apr. | Hitler authorises the private purchase of sabres by army officers |
| 1 May | (EAGLE AND SWASTIKA EMBLEM OF THE *NSDAP* IS INCORPORATED INTO GERMAN ARMY UNIFORM) |
| 1 May | Nazi eagle replaces that of the Weimar Republic on *Waffenamt* inspection stamps |
| 30 Jun. | (NIGHT OF THE LONG KNIVES PURGE ELIMINATES THE *SA* CHALLENGE TO HITLER'S AUTHORITY) |
| 3 Jul. | *SS* Himmler Honour Dagger introduced |
| 4 Jul. | *SA* Röhm Honour Daggers recalled |
| 4 Jul. | *SS* Röhm Honour Daggers recalled |
| 4 Jul. | *SA-Stabswache Göring* sabre discontinued |
| 2 Aug. | (HITLER BECOMES HEAD OF STATE AND CHIEF OF THE ARMED FORCES) |
| 30 Aug. | Naval *SA* dagger introduced |

| | |
|---|---|
| 1 Oct. | (HITLER ORDERS EXPANSION OF THE ARMY AND NAVY, AND CREATION OF A NEW *LUFTWAFFE*) |

– 1935 –

| | |
|---|---|
| 25 Jan. | *SS* members dismissed from the organisation are ordered to surrender their daggers |
| 26 Feb. | *DLV* sword adopted by the *Luftwaffe* |
| 1 Mar. | Production of *DLV* dagger discontinued |
| 1 Mar. | *Luftwaffe* airman's dagger introduced |
| 1 Mar. | *Luftwaffe* general's sword introduced |
| 16 Mar. | (CONSCRIPTION REINTRODUCED IN GERMANY) |
| 16 Mar. | System of *RZM* codes formalised |
| 1 Apr. | *SA-Feldjägerkorps* sabre discontinued |
| 1 Apr. | *SA-Feldjägerkorps* bayonet discontinued |
| 4 May | Army dagger introduced |
| 18 Jun. | Publication of order covering the wearing of the Army dagger and its accoutrements |
| 21 Jun. | *SA* Honour Dagger introduced |
| 15 Sep. | (SWASTIKA BECOMES THE OFFICIAL NATIONAL EMBLEM OF THE GERMAN STATE) |
| 9 Nov. | *NPEA* student's dagger introduced |
| 9 Nov. | *NPEA* staff dagger introduced |

– 1936 –

| | |
|---|---|
| 7 Mar. | (GERMANY REOCCUPIES THE RHINELAND) |
| 22 Mar. | Hunting Association cutlass introduced |
| 20 Apr. | *RZM* assumes control of the production and supply of *SA*, *SS* and Hitler Youth daggers and knives |
| 20 Apr. | Naval *SA* dagger discontinued |
| 19 May | Black scabbard authorised for the *NSKK* dagger, to distinguish it from that of the *SA* |
| 27 May | Fire Brigade dagger discontinued |
| 27 May | Fire Brigade axe discontinued |
| 27 May | Fire Brigade bayonet introduced |
| 1 Jun. | Aluminium replaces nickel silver in the manufacture of the *Luftwaffe* sword and airman's dagger |

| | |
|---|---|
| 21 Jun. | *SS* chained dagger introduced |
| 21 Jun. | *SS* Honour Dagger introduced |
| 21 Jun | *SS* officer's sword introduced |
| 21 Jun. | *SS* NCO's sword introduced |
| 21 Jun. | Police officer's sword introduced |
| 21 Jun. | Police NCO's sword introduced |
| 21 Jun. | Design of Police bayonet revised |
| 1 Jul. | Customs Service bayonet introduced |
| 5 Jul. | *RLB* bayonet introduced |
| 5 Jul. | *RLB* knife introduced |
| 5 Jul. | *RLB* dagger introduced |
| 2 Aug. | *NSKK* chained dagger introduced |
| 10 Aug. | Justice Service sabre introduced |
| 10 Aug. | Prison Service sabre introduced |
| 15 Sep. | *NPEA* chained dagger introduced |

– 1937 –

| | |
|---|---|
| 19 Jan. | Use of Army dagger extended to officer cadets |
| 17 Apr. | *DLV* knife adopted by the *NSFK* |
| 20 Apr. | Hitler Youth leader's dagger introduced |
| 20 Apr. | First Damascus-bladed *SS* sword presented to Adolf Hitler |
| 10 Jul. | *SA Feldherrnhalle* dagger introduced |
| 31 Jul. | Land Customs dagger introduced |
| 31 Jul. | Water Customs dagger introduced |
| 25 Sep. | (STATE VISIT OF MUSSOLINI TO BERLIN) |
| 1 Oct. | *Luftwaffe* officer's dagger introduced |
| 1 Oct. | Design of *Luftwaffe* general's sword revised |
| 3 Dec. | *RAD* leader's hewer introduced |
| 28 Dec. | Golden *SA Feldherrnhalle* dagger presented to *SA-Stabschef* Lutze |

– 1938 –

| | |
|---|---|
| 12 Jan. | Golden *SA Feldherrnhalle* dagger presented to Göring |
| 2 Feb. | *SA* chained Honour Dagger introduced |
| 28 Feb. | Red Cross hewer introduced |
| 28 Feb. | Red Cross dagger introduced |
| 1 Mar. | *NSKK* Honour Dagger introduced |
| 12 Mar. | (GERMANY OCCUPIES AUSTRIA) |

| | |
|---|---|
| 9 Apr. | Railway Guard dagger introduced |
| 20 Apr. | Design of Navy dagger revised |
| 20 Apr. | Navy Honour Dagger introduced |
| 20 Apr. | Swastika incorporated into blade etchings on Navy sabres |
| 20 Apr. | Waterways Protection Police dagger introduced |
| 30 Apr. | Rail Waterways Protection Police dagger introduced |
| 1 May | Diplomatic Corps dagger introduced |
| 1 May | Diplomatic Corps sword introduced |
| 1 May | Diplomatic Corps bayonet introduced |
| 20 Jun. | Forestry Service cutlass introduced |
| 10 Jul. | Fire Brigade officers authorised to wear the Police officer's sword |
| 28 Aug. | Blade motto on Hitler Youth knife discontinued |
| 29 Aug. | Design of *RLB* bayonet revised |
| 29 Aug. | Design of *RLB* knife revised |
| 29 Aug. | Design of *RLB* dagger revised |
| 1 Oct. | (GERMANY OCCUPIES THE SUDETENLAND) |
| 30 Nov. | *Teno* hewer introduced |
| 30 Nov. | *Teno* dagger introduced |

### – 1939 –

| | |
|---|---|
| 1 Feb. | Postal Protection Corps dagger introduced |
| 8 Feb. | Rifle Association cutlass introduced |
| 15 Mar. | (GERMANY OCCUPIES CZECHOSLOVAKIA) |
| 30 Mar. | Government administration dagger introduced |
| 1 Apr. | Production of Waterways Protection Police dagger discontinued |
| 22 May | (GERMAN–ITALIAN PACT) |
| 1 Jul. | Flight utility knife introduced |
| 23 Aug. | (NAZI–SOVIET PACT) |
| 1 Sep. | (GERMANY INVADES POLAND, CAUSING GREAT BRITAIN AND FRANCE TO DECLARE WAR) |
| 27 Sep. | (POLAND SURRENDERS) |
| 31 Dec. | First Navy Honour Dagger presented to *Vizeadmiral* Albrecht |
| 31 Dec. | Production of *RLB* bayonet discontinued |

### – 1940 –

| | |
|---|---|
| 1 Jan. | Production of Police officer's sword discontinued |
| 1 Jan. | Production of Police NCO's sword discontinued |
| 7 Feb. | Production of Rifle Association cutlass discontinued |
| 14 Feb. | Government General dagger introduced |
| 12 Mar. | Use of *Luftwaffe* officer's dagger extended to senior NCOs |
| 5 Apr. | Production of Red Cross hewer discontinued |
| 5 Apr. | Production of Red Cross dagger discontinued |
| 9 Apr. | (GERMANY INVADES DENMARK AND NORWAY) |
| 10 May | (GERMANY INVADES THE LOW COUNTRIES) |
| 24 May | *Waffen-SS* dagger proposed |
| 22 Jun. | (GERMANY VICTORIOUS IN THE WEST) |
| 19 Jul. | *Reichsmarschall* dagger presented to Göring |
| 18 Aug. | Fisheries Control Department dagger introduced |
| 1 Sep. | Production of *SS* 1933 dagger discontinued |
| 17 Sep. | (HITLER POSTPONES INVASION OF GREAT BRITAIN) |
| 20 Dec. | Production of Army ordnance sabre discontinued |
| 28 Dec. | First Army Honour Dagger presented to *SA-Stabschef* Lutze |

### – 1941 –

| | |
|---|---|
| 25 Jan. | Production of *SS* officer's sword discontinued |
| 25 Jan. | Production of *SS* NCO's sword discontinued |
| 23 Feb. | Railway Service dagger introduced |
| 19 Mar. | Production of Railway Guard dagger discontinued |
| 19 Mar. | Production of Rail Waterways Protection Police dagger discontinued |
| 10 Apr. | Hitler Youth *Wachgefolgschaft* bayonet introduced |
| 22 Jun. | (GERMANY INVADES THE SOVIET UNION) |
| 10 Aug. | Production of Hitler Youth knives reaches 15 million |

| | |
|---|---|
| 10 Aug. | Production of Army private-purchase sabres reaches 5 million |
| 5 Dec. | (GERMANS ABANDON ASSAULT ON MOSCOW) |
| 11 Dec. | (GERMANY DECLARES WAR ON THE USA) |
| 31 Dec. | Production of *Wehrmacht* dress bayonet discontinued |
| 31 Dec. | Production of Police bayonet discontinued |
| 31 Dec. | Production of Customs Service bayonet discontinued |
| 31 Dec. | Production of Justice Service sabre discontinued |
| 31 Dec. | Production of Prison Service sabre discontinued |

## – 1942 –

| | |
|---|---|
| 17 Jan. | Synthetic Celleon replaces aluminium bullion wire in the manufacture of dagger portepée knots |
| 5 Feb. | (WEHRMACHT LOSSES REACH 1 MILLION) |
| 25 Mar. | Eastern Ministry dagger introduced |
| 15 Apr. | (HEAVY BOMBING OF GERMAN CITIES BEGINS) |
| 4 May | Revised manner of wear prescribed for the Army dagger knot, to conserve material |
| 31 May | Gold fittings introduced for the dagger hanging straps worn by *Luftwaffe* general officers |
| 15 Jul. | Production of Diplomatic Corps, government administration, Government General, Land Customs, Water Customs, *RLB*, Forestry Service and Hunting Association daggers, swords, bayonets, knives and cutlasses discontinued |
| 10 Oct. | Gold fittings introduced for the dagger hanging straps worn by Army general officers |
| 14 Oct. | Production of *NPEA* and Hitler Youth daggers and knives discontinued |
| 2 Nov. | (BRITISH OFFENSIVE OPENS IN NORTH AFRICA) |
| 8 Nov. | Further wear of Diplomatic Corps, government administration and Government General daggers prohibited |
| 19 Nov. | (BATTLE OF STALINGRAD COMMENCES) |

## – 1943 –

| | |
|---|---|
| 2 Feb. | (GERMANS SURRENDER AT STALINGRAD) |
| 15 Feb. | *SS* chained dagger and knot authorised to be worn with field-grey service dress by *Waffen-SS* officers |
| 15 Feb. | Direct sale of *SS* chained dagger via *RZM*-approved outlets permitted |
| 18 Feb. | (GOEBBELS MOBILISES THE GERMAN CIVILIAN POPULATION FOR 'TOTAL WAR') |
| 25 Mar. | Production of *Luftwaffe* sword discontinued |
| 1 Apr. | Production of all edged weapon knots discontinued |
| 13 May | (GERMAN FORCES IN NORTH AFRICA CAPITULATE) |
| 27 May | Further production of most remaining dress daggers, swords and sabres discontinued. Only Navy daggers, *SS* chained daggers, *NSFK* knives and special presentation items are not covered by the cancellation order |
| 4 Jun. | *SS* chained dagger and knot authorised to be worn with field-grey service dress by Security Police and *SD* officers |
| 26 Aug. | (LONG GERMAN RETREAT ON THE EASTERN FRONT BEGINS) |
| 31 Dec. | Production of *SS* chained dagger discontinued |

## – 1944 –

| | |
|---|---|
| 10 Feb. | Production of *NSFK* knife discontinued |
| 25 Feb. | Wearing of the Navy dagger with field-grey marine uniform prohibited |
| 9 May | Final presentation of a Navy Honour Dagger to *Fregattenkapitän* Brandi |
| 10 May | Production of Navy dagger discontinued |
| 16 May | (PEAK OF GERMAN ARMAMENTS PRODUCTION) |
| 16 May | Solingen blade manufacture now devoted almost entirely to *Wehrmacht* service bayonets and fighting knives |
| 6 Jun. | (D-DAY LANDINGS IN NORMANDY) |
| Aug.–Sep. | Solingen subjected to sustained area bombing by the British and American air forces, |

with over 40% of the city destroyed: edged weapons production severely curtailed as a result

12 Sep.  (US ARMY REACHES THE GERMAN BOR-DER)

23 Dec.  Further wearing of all *Wehrmacht* dress daggers and swords prohibited, in favour of pistols

– 1945 –

3 Mar.  (US ARMY CROSSES THE RHINE)

1 Apr.  (RUHR POCKET ENCIRCLED BY US 1st AND 9th ARMIES)

18 Apr.  Solingen captured by the Americans

23 Apr.  (RUSSIANS ENTER BERLIN)

30 Apr.  (HITLER COMMITS SUICIDE)

7 May  (GENERAL JODL SIGNS GERMAN SURREN-DER)

23 May  (DÖNITZ GOVERNMENT DISBANDED: THIRD REICH OFFICIALLY ENDS)

23 May – end of year    Allied forces in Germany systematically gather surviving Nazi edged weapons for mass destruction

# APPENDIX II:

# WHERE TO OBTAIN NAZI EDGED WEAPONS

Third Reich daggers, swords, bayonets and knives have been popular with militaria collectors since 1945. They can be bought from specialist dealers, at auction or by attending the arms fairs regularly held at weekends throughout the country. Some relevant contact addresses are shown below. However, it must be borne in mind that there are many fake Nazi edged weapons around and caution must always be exercised.

## DEALERS

A. Beadle
PO Box 1658
Dorchester
DT2 9YD
Tel. 01308 897904

Blunderbuss Antiques
29 Thayer Street
London
W1M 5LT
Tel. 020 7486 2444

Chelsea Military Antiques
Unit N13/14, Antiquarius
131–141 Kings Road
London
SW3 4PW
Tel. 020 7352 0308

M. Coverdale
Glenwood
210 Darlington Lane
Stockton-on-Tees
TS19 8AD
Tel. 01642 603627

J. Cross
PO Box 73
Newmarket
Suffolk
CB8 2RY
Tel. 01638 750132

A. Forman
PO Box 163
Braunton, EX33 2YF
Tel. 01271 816177

C. James
Medals and Militaria
Warwick Antiques Centre
22–24 High Street
Warwick
CV34 4AP
Tel. 01926 495704

Johnson Reference Books & Militaria
312 Butler Road
Chatham Square Office Park No. 403
Fredericksburg
Virginia 22405
USA
Tel. 540 373 9150

Just Military
701 Abbeydale Road
Sheffield
S7 2BE
Tel. 0114 255 0536

M. & T. Militaria
The Banks
Bank Lane
Victoria Road
Carlisle
CA1 2UA
Tel. 01228 530988

Military Antiques
11 The Mall Antiques Arcade
359 Upper Street
Islington
London
N1 0PD
Tel. 020 7359 2224

Mons Military Antiques
221 Rainham Road
Rainham
Essex
RM13 7SD
Tel. 01277 810558

The Old Brigade
10a, Harborough Road
Kingsthorpe
Northampton
NN2 7AZ
Tel. 01604 719389

Platoon
77–79 Chapel Street
Salford
Manchester
M3 5BZ
Tel. 0161 839 5185

Regimentals
PO Box 130
Hitchin
Herts
SG5 2BE
Tel. 01462 713294

The Treasure Bunker
21 King Street
Glasgow
G1 5QZ
Tel. 0141 552 4651

Ulric of England
PO Box 285
Stoneleigh
Epsom
Surrey
KT17 2YJ
Tel. 020 8393 1434

Wittmann Militaria
1253 North Church Street
Moorestown
New Jersey 08057
USA
Tel. 609 235 0622

# AUCTION HOUSES

Bosleys Military Auctioneers
42 West Street
Marlow
Buckinghamshire
SL7 2NB
Tel. 01628 488188

Kent Sales
The Street
Horton Kirby
Dartford
DA4 9BY
Tel. 01322 864919

Wallis & Wallis
West Street Auction Galleries
Lewes
Sussex
BN7 2NJ
Tel. 01273 480208

Warwick & Warwick
Chalon House
Scar Bank
Millers Road
Warwick
CV34 5DB
Tel. 01926 499031

# ARMS FAIRS

A.J.W. Militaria
PO Box HP96
Leeds
L56 3XU
Tel. 0113 275 8060

Antique Militaria Exhibitions
PO Box 104
Warwick
CV34 5ZG
Tel. 01926 497340

Arms & Armour UK
58 Harpur Street
Bedford
MK40 2QT
Tel. 01234 344831

H&S Militaria Fairs
PO Box 254
Tonbridge
Kent
TN12 7ZQ
Tel. 01892 730233

# APPENDIX III:
# PRICE GUIDE

The market value of Third Reich edged weapons can vary quite significantly depending upon a number of factors, primarily whether pieces are being bought or sold and whether dealers or auction houses are involved. In general terms dealers seek to make at least 30 per cent profit on any given transaction, while auctioneers charge customers a range of commissions to allow for their percentage and overheads. Items can normally be obtained more cheaply at auction, but two determined bidders may easily raise the final hammer price to one which is well above the lot's catalogue estimate. Condition, quality and the presence or otherwise of an accompanying portepée knot or hanging straps can also have a dramatic effect on desirability and cost. Consequently, as in every field of collecting, there can be no such thing as a 'set value' for any given Nazi dagger, sword, bayonet or knife.

This section is intended to provide a general guide to the average prices which collectors might expect to pay if purchasing original pieces in good condition from reputable dealers, based on the market at the time of publication. Scarce and highly sought-after Third Reich edged weapons have consistently risen in value over the decades, a typical example being the *SS* chained dagger, which has seen the following price increases:

1970 – £200
1980 – £650
1990 – £1500
2000 – £2300

However, more common items like the Army dagger and the Hitler Youth knife have experienced far more moderate rises. In short, the normal rules of

**181**. *The presence of rare accoutrements, such as the* Bahnschutz *knot and straps shown here, can greatly affect the price of an edged weapon.*

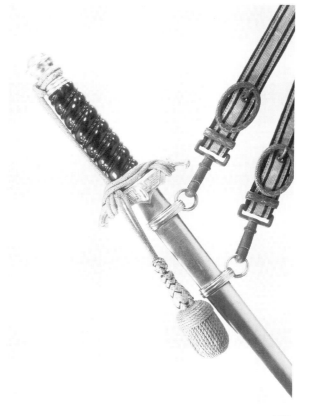

'supply and demand' hold good. Unique and exceptionally rare presentation items seldom appear on the market, so are shown as having 'speculative' value.

| DAGGERS | £ |
|---|---|
| Army 1935 dagger | 280 |
| Army 1935 dagger with etched blade | 700 |
| Army 1935 dagger with etched artificial Damascus blade | 1000 |
| Army 1935 dagger with Damascus blade | 1500 |
| Army 1940 Honour Dagger | Speculative |
| Navy 1938 dagger | 480 |
| Navy 1938 Honour Dagger | Speculative |
| *Luftwaffe* 1935 dagger | 450 |
| *Luftwaffe* 1937 dagger | 350 |
| *SA* 1933 dagger | 280 |
| *SA* 1934 Röhm Honour Dagger with complete dedication | 1250 |
| *SA* 1934 Röhm Honour Dagger with partial dedication | 680 |
| *SA* 1934 Röhm Honour Dagger with erased dedication | 300 |
| *SA* 1935 Honour Dagger | Speculative |
| *SA* 1938 Honour Dagger | Speculative |
| *SA Feldherrnhalle* dagger | Speculative |
| Naval *SA* dagger | 1500 |
| *SS* 1933 dagger | 1000 |
| *SS* 1934 Röhm Honour Dagger | Speculative |
| *SS* 1934 Himmler Honour Dagger | Speculative |
| *SS* 1936 chained dagger | 2300 |
| *SS* 1936 Honour Dagger | Speculative |
| *NSKK* 1933 dagger | 300 |
| *NSKK* 1936 chained dagger | 1800 |
| Naval *NSKK* 1936 chained dagger | 3000 |
| *NSKK* 1938 Honour Dagger | Speculative |
| *NSKK* High leader's dagger | Speculative |
| *NPEA* 1935 student's dagger | 1600 |
| *NPEA* 1935 staff dagger | 2000 |
| *NPEA* 1936 staff chained dagger | 3000 |
| *HJ* 1933 knife with blade motto | 170 |
| *HJ* 1933 knife without blade motto | 130 |
| *HJ* leader's 1937 dagger | 2250 |
| *DLV* 1934 dagger | Speculative |
| *DLV* 1934 knife | 850 |
| *NSFK* 1937 knife | 550 |
| Diplomatic Corps 1938 dagger | 2500 |
| Government administration 1939 dagger | 2300 |
| Fisheries Control Department 1940 dagger | Speculative |
| Government General 1940 dagger | Speculative |
| Waterways Protection Police 1938 dagger | 1500 |
| Fire Brigade 1870 dagger | 750 |
| Fire Brigade 1870 axe | 1250 |
| Land Customs 1937 dagger | 850 |
| Water Customs 1937 dagger | 3000 |
| Postal Protection Corps 1939 dagger | 2250 |
| Railway Guard 1938 dagger | 1900 |
| Rail Waterways Protection Police 1938 dagger | Speculative |
| Railway Service 1941 dagger | Speculative |
| *RAD* 1934 hewer | 450 |
| *RAD* Leader's 1937 hewer | 800 |
| *RLB* 1936 knife | 1100 |
| *RLB* 1936 dagger | 1900 |
| *RLB* 1938 knife | 800 |
| *RLB* 1938 dagger | 1750 |
| *Teno* 1938 hewer | 1650 |
| *Teno* 1938 dagger | 2500 |
| *DRK* 1938 hewer | 250 |
| *DRK* 1938 dagger | 550 |
| Forestry Service 1938 cutlass | 650 |
| Hunting Association 1936 cutlass | 750 |
| Rifle Association 1939 cutlass | 1100 |

| SWORDS | £ |
|---|---|
| Army enlisted man's ordnance sabre | 175 |
| Army officer's ordnance sabre | 250 |
| Army lion-head sabre | 325 |
| Army lion-head sabre with etched blade | 850 |
| Navy sabre | 750 |
| *Luftwaffe* 1934 sword | 550 |
| *Luftwaffe* general's 1935 sword | Speculative |
| *Luftwaffe* general's 1937 sword | Speculative |
| *SA-Stabswache* Göring 1933 sabre | Speculative |
| *SA-Feldjägerkorps* 1933 sabre | Speculative |
| *SS* lion-head sabre | 2250 |
| *SS* officer's 1936 sword | 3000 |

| | |
|---|---|
| *SS* NCO's 1936 sword | 850 |
| *SS* birthday sword | Speculative |
| Diplomatic Corps 1938 sword | Speculative |
| Police officer's 1936 sword | 650 |
| Police NCO's 1936 sword | 500 |
| Water Customs sabre (Eickhorn model) | 2300 |
| Justice Service 1936 sabre | 1600 |
| Prison Service 1936 sabre | 850 |
| Miner's sabre | 500 |
| Rifle Association 1936 sword | 650 |
| Presentation swords | Speculative |

| BAYONETS | £ |
|---|---|
| *Wehrmacht* M.84/98 service bayonet | 50 |
| *Wehrmacht* M.98 dress bayonet with plain blade | 85 |
| *Wehrmacht* M.98 dress bayonet with Army commemorative etched blade | 220 |
| *Wehrmacht* M.98 dress bayonet with *Luftwaffe* commemorative etched blade | 350 |
| *Wehrmacht* presentation Bayonet of Honour | 850 |
| Hitler Youth *Wachgefolgschaft* Bayonet | 350 |
| Diplomatic Corps bayonet | Speculative |
| *SA-Feldjägerkorps* 1933 bayonet | 900 |
| Police M.26 long clamshell bayonet | 750 |
| Police M.36 long service bayonet | 280 |
| Police M.36 short dress bayonet | 450 |
| Fire Brigade bayonet | 120 |
| Customs Service bayonet | 800 |
| *RLB* bayonet | 350 |
| *Stahlhelm* bayonet | 300 |
| *NSKOV* bayonet | 400 |

| KNIVES | £ |
|---|---|
| Standard fighting knife | 90 |
| *Waffen-Loesche* knife | 250 |
| Flight utility knife | 280 |
| *Luftwaffe* survival machete | 450 |
| Diver's knife | 900 |
| M.44 combination knife | 350 |

| HANGERS AND STRAPS | £ |
|---|---|
| Hanging straps for Army dagger | 80 |
| Hanging straps for Navy dagger | 300 |
| Hanging chains for *Luftwaffe* 1935 dagger | 120 |
| Hanging straps for *Luftwaffe* 1937 dagger | 130 |
| Hanging strap for *SA* 1933 dagger | 40 |
| Belt loop for *SA* 1933 dagger hanging strap | 30 |
| Three-piece vertical hanger for *SA* 1933 dagger | 90 |
| Hanging strap for *SS* 1933 dagger | 50 |
| Belt loop for *SS* 1933 dagger hanging strap | 40 |
| Vertical hanger for *SS* 1933 dagger | 150 |
| Frog for *NPEA* student's dagger | 280 |
| Hanging straps for Land Customs dagger | 220 |
| Leather hanging straps for *RAD* leader's 1937 hewer | 450 |
| Leather hanging straps for *RLB* 1938 dagger | 400 |
| Frog for *Teno* hewer | 350 |
| Leather hanging straps for *Teno* dagger | 1000 |
| Brocade hanging straps for *Teno* dagger | 1250 |
| Frog for *DRK* hewer | 90 |
| Medical/First Aid hanging straps for *DRK* dagger | 250 |
| Social Welfare hanging straps for *DRK* dagger | 380 |
| Frog for Hunting Association cutlass | 100 |
| Hanger for Army sabre | 50 |
| Shoulder harness for Army sabre | 120 |
| Hanger for *Luftwaffe* sword | 90 |
| Leather hanger for *SS* officer's 1936 sword | 180 |
| Brocade hanger for *SS* officer's 1936 sword | 500 |
| Frog for *Wehrmacht* M.84/98 service bayonet | 30 |
| Frog for *Wehrmacht* M.98 dress bayonet | 40 |

| KNOTS | £ |
|---|---|
| Army dagger knot | 50 |
| Navy dagger gold knot | 90 |
| *Luftwaffe* dagger knot | 60 |
| Diplomatic Corps dagger knot | 450 |

| | |
|---|---|
| Railway Guard dagger knot | 500 |
| Forestry Service cutlass knot | 150 |
| Rifle Association cutlass knot | 120 |
| Army officer's sabre knot | 40 |
| *SS* officer's sword knot | 400 |
| *SS* NCO's sword knot | 500 |
| Police sword knot | 50 |
| Land Customs sabre knot | 220 |
| Prison Service sabre knot | 250 |
| Army bayonet knot | 30 |

| SPARE PARTS | £ |
|---|---|
| Army dagger blade | 70 |
| Army dagger pommel | 25 |
| Army dagger grip | 60 |
| Army dagger scabbard | 110 |
| Navy dagger Damascus blade | 600 |
| *SA* dagger blade | 90 |
| *SA* dagger grip eagle | 10 |
| *SA* dagger scabbard | 90 |
| *SA* chained Honour Dagger scabbard | 3000 |
| *SS* dagger blade | 350 |
| *SS* dagger grip runes | 50 |
| *SS* dagger scabbard | 130 |
| *HJ* knife grip plates | 20 |
| *HJ* leader's dagger pommel | 150 |
| Government administration dagger crossguard | 250 |
| Postal Protection Corps dagger grip | 250 |
| *DRK* hewer blade | 80 |
| *DRK* dagger blade | 150 |
| Army sabre blade | 90 |
| Army sabre grip | 60 |

| | |
|---|---|
| Army sabre scabbard | 90 |
| *Luftwaffe* sword blade | 150 |
| *Luftwaffe* sword pommel | 50 |
| *SS* officer's sword grip runes | 200 |
| Police officer's sword grip eagle | 40 |
| Diplomatic Corps sword hilt fittings | 850 |
| Police bayonet staghorn grip plates | 30 |

| MINIATURES | £ |
|---|---|
| Miniature Army dagger | 380 |
| Miniature Navy dagger | 550 |
| Miniature *Luftwaffe* 1935 dagger | 450 |
| Miniature *Luftwaffe* 1937 dagger | 400 |
| Miniature *SS* chained dagger | Speculative |
| Miniature *RAD* leader's hewer | 1500 |
| Miniature Hunting Association cutlass | 900 |
| Miniature *Luftwaffe* sword | 600 |
| Miniature *SS* Officer's Sword | Speculative |
| Miniature *Wehrmacht* dress bayonet | 250 |

| SALES MATERIAL | £ |
|---|---|
| Edged weapon maker's sales catalogue | 250 |
| Edged weapon maker's metal advertising sign | 350 |
| Edged weapon maker's calendar | 120 |
| Factory guarantee tag | 40 |
| Dagger paper storage bag | 60 |
| Sword cloth carrying bag | 170 |
| Copy of *Die Klinge* | 30 |

# APPENDIX IV:

# POST-1945 REPRODUCTIONS

Having discussed where to buy Third Reich edged weapons and what to pay for them, we need to consider the thorny issue of faking, as many Nazi bladed sidearms have been copied since 1945 to satisfy the growing and ever more lucrative collectors' market. Reproductions fall neatly into the five main categories of daggers, swords, bayonets, knives and accoutrements.

## DAGGERS

Dress daggers have been by far the most extensively faked of all Third Reich edged weapons, with scores of variants in a range of qualities being made after the end of the Second World War. Just as original pieces such as the *NPEA* 1936 staff chained dagger, the Waterways Protection Police and Rail Waterways Protection Police 1938 daggers, and the Postal Protection Corps 1939 dagger were created at short notice using readily available parts from existing dagger patterns for reasons of economy, so a number of modern copies have been dismantled and reassembled, employing components from originals and other fakes. This has resulted in an almost indescribable array of hybrid 'parts daggers' and a good deal of confusion into the bargain. The following subsections cover the features common to known reproductions of particular types of dagger and should provide a fairly comprehensive guide to what to look out for, and beware of, when examining any given piece.

### THE ARMY 1935 DAGGER

The Army dagger was produced in many variations by dozens of different firms between 1935 and 1943. Early examples were manufactured with nickel silver or silver-plated bronze fittings and a plated steel scabbard, while later pieces featured whitened zinc components. The handles could be white, yellow, orange-red or black in colour. Most grips were solid plastic or plastic over a wood base, while a few de luxe ones were in ivory. Crossguard eagles varied slightly in design, certain steel blades were purer and less prone to rust than others, and so on.

The differences between originals can therefore be considerable. Nevertheless, most copies are still readily identifiable as such since fakes usually have one or more of the following characteristics:

- poor definition – originals always had good detailing to the eagle, oakleaves and scabbard pebbling
- brass fittings – genuine Army daggers were not constructed from brass
- grip grooves running in the wrong direction, i.e. from high left to low right – they should run from high right to low left
- grips with very sharp ridges to the grooves – originals had smooth curves
- *RZM* stamps to the blade – Army blades were never *RZM* marked
- rounded edges to the flats of the blade – these should be sharply defined

- scabbards with separate tips and a distinctly swirling pattern to the pebbling – original scabbards were made in one piece, with dense and irregular pebbling
- narrow brass liners soldered into the scabbard – genuine liners were in steel or zinc alloy and were riveted into position.

A variation Army dagger was considered in 1941, and was illustrated in *Blanke Waffen* along with the proposed Field Marshal's Sword and stainless steel sabres. However, it never progressed beyond the prototype stage. The design for the new dagger featured a vertically grooved white grip, a pommel bearing *Wehrmacht* eagles on the obverse and reverse, and a spiralling scroll to the crossguard. Many copies of it now circulate, most with the mark of Ernst Pack etched on their blades. Some display odd additions to their pommels, namely two triangular shields each containing a Sig-rune of the type encountered on the *SS* death's head ring, and these weapons have been offered spuriously as rare Honour Daggers presented by Heinrich Himmler to senior *Waffen-SS* officers.

## THE NAVY 1938 DAGGER

The Navy dagger was produced in even more variations than its Army counterpart. For instance, the purchaser could opt for a plain or an etched blade, and have either a hammered or an engraved scabbard. Grips might be white or yellow, and of horn, ivory, solid plastic or plastic over wood. Even the suspension rings could be plain or of a rope or oakleaf design, differing from maker to maker. But collectors will find that the majority of genuine naval sidearms encountered feature an etched blade and a scabbard engraved with lightning bolts.

Marine daggers are still produced in Germany for export and wear by domestic naval personnel. The pommel eagle now holds a fouled anchor rather than a swastika in its talons, but otherwise the overall design is virtually unchanged. As a result, many modern daggers have simply had their pommels replaced by Nazi ones and been passed off as wartime issues. Certain indicators do exist, howev-

er, to assist the collector in differentiating between an authentic Third Reich dagger and a touched-up postwar piece. These can be summarised as follows:

- loose pommels – replacements are usually slack or off-centre when screwed into position
- thin wire wrapping to the grip – original wire wrapping was finely made but substantial
- gilded steel or lightweight alloy fittings and scabbards – instead of the original gold-plated brass
- plastic scabbard liners – never used during the Third Reich
- incorrect etching or engraving – blades etched with a fouled anchor surmounted by a *Wehrmacht* eagle, or with a modern battleship, or with the Kiel naval monument, are post-war creations as these designs were not known prior to 1945, and the same applies to scabbards engraved with battleships or sea serpents
- incorrect makers – only seventeen were authorised to produce naval daggers. Blades bearing other trademarks are post-1945 productions.

## THE *LUFTWAFFE* 1935 DAGGER

The *Luftwaffe* 1935 dagger, with its grip and scabbard wrapped in fine-grain leather, was an expensive item to manufacture. Early examples had the metallic portions in nickel silver with small connecting rings on the chain hangers, while post-1936 pieces were distinguished by their more economical aluminium fittings and larger connecting rings. Reproductions can be identified by one or more of the following features:

- deformed sunwheel swastikas on the pommel and crossguard – originals were symmetrically perfect
- grip wire of a single double strand – the authentic arrangement was of two double strands lying alongside each other, or of a double strand set between two single strands
- ill-shaped grips
- badly attached leather or PVC wrappings to the grip and scabbard
- upper scabbard mounts of excessive length – the

original left a space of about one and a half times its own length between the locket and the scabbard's central mount

- crude liners soldered into the scabbard – these should be well formed and riveted into position
- rounded edges to the flats of the blade – genuine blades had sharply defined edges
- *RZM* markings to the blade – *Luftwaffe* blades were never *RZM* stamped
- a spurious stamped representation of the Clemen & Jung trademark, being an oval cartouche containing the words 'Clemen & Jung, Solingen' around a 'Z' within a shield – the correct Third Reich version of the mark was simply a 'Z' within a shield

### THE *LUFTWAFFE* 1937 DAGGER

The second pattern *Luftwaffe* dagger, like its Army and Navy counterparts, came in many variations. Early examples bore hilt fittings of white cast aluminium while later issues had zinc-based alloy components with a dark grey finish. Some pommels had gilded swastikas while others had grey ones. Grips could be white or yellow and of ivory, solid plastic or plastic over wood. The form of the crossguard eagle, particularly in the fletching of its wings, varied from maker to maker and blades might be plain or etched. The only common denominator appears to have been the scabbard, which was of grey steel in all cases.

The following characteristics are common to reproductions:

- crossguards in a heavy grey alloy – original examples, including the zinc types, were always lightweight
- grip wire running in the wrong direction, i.e. from high right to low left – it should run from high left to low right (note how this arrangement differs from the grooves on Army and Navy handles)
- grip wire comprising a twisted double strand – the original wire was a single strand, thick and wavy in form
- *RZM* marks to the blade – *Luftwaffe* blades were never *RZM* stamped
- scabbards with separate tips and a distinctive

swirling pattern to the pebbling – genuine scabbards were constructed in one piece, with dense and irregular pebbling
- scabbards in a grey or silver-coloured non-magnetic alloy as opposed to the steel originals
- poor stippling behind the oakleaves on the shield at the lower end of the scabbard – this should be well defined
- wooden or brass liners crudely fitted into the scabbard locket – genuine liners were in steel or zinc alloy and were riveted into position

Regulation political and civil daggers were more standardised than those of the *Wehrmacht* and less prone to contemporary variation. The following subsections describe the salient features common to reproductions of each particular type.

### THE *SA* 1933 DAGGER

- plastic grips – originals were in a variety of woods, especially walnut and maple, carved symmetrically, stained and varnished
- a poorly cast eagle set too high on the grip – the top of the eagle should be level with the widest part of the grip

**182.** SA *grip eagles. The example on the left is original, while that on the right is a fake.*

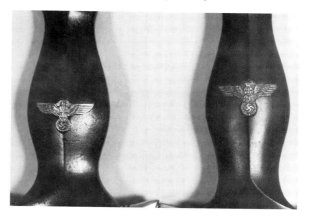

- badly etched blade mottoes with ragged edging, a stippled background to the lettering, or the wording misaligned – etching should be crisp, smooth and level
- 'Germany' stamped on the blade ricasso or tang – indicative of a postwar dagger made for export
- sharp blade spines – genuine blades had rounded spines
- poorly finished scabbards with misshapen rims to the mounts
- cast lockets and chapes with prominent edge seams – originals were die-struck
- one-piece lockets – originals were in two sections
- thin scabbard liners soldered or simply pushed into place inside the locket –original liners were wide, of brass, and riveted into position

## THE SA 1934 RÖHM HONOUR DAGGER

As for the SA 1933 dagger, but with the following additional features:

- poorly etched Röhm dedications – presentation etching was always of excellent quality
- engraved Röhm dedications – originals were never engraved
- Damascus or artificial Damascus Röhm blades with raised gilded lettering – never a feature of originals
- offset dedications – the dedication should be central along the reverse spine
- a maker other than Böker & Co., Carl Eickhorn, C.G. Haenal, Gottlieb Hammesfahr, J.A. Henckels, Richard Herder, Ernst Pack, Anton Wingen or Eduard Wüsthof, the only known original producers of Röhm blades
- RZM-marked blades – these were not introduced until after the Röhm Honour Dagger was withdrawn from production

## THE SA 1935 HONOUR DAGGER

As for the SA 1933 dagger, but with these additional characteristics:

- badly cast grip mounts in copper, brass or soft alloy with edge seams – originals were always die-struck in nickel silver, with fine detailing to the oakleaves and acorns
- engraved grip mounts – never encountered on originals
- blades bearing non-Eickhorn trademarks – only Carl Eickhorn was authorised to produce the SA 1935 Honour Dagger
- RZM-marked blades – Honour Daggers were never RZM marked

## THE SA 1938 HONOUR DAGGER

As for the SA 1935 Honour Dagger, but with the following additional points:

- faint background stippling to the suspension chain links – this stippling should not be present
- circular junction rings to the chain links – authentic examples were semicircular
- scabbards decorated with runic emblems, in particular sunwheel swastikas, life runes and Tyr-runes – never a feature of originals

## THE SA FELDHERRNHALLE DAGGER

- pommel locking nuts with twin holes – originals had cruciform slots
- grips in soft wood – genuine examples were in hard wood
- wooden scabbard liners – these should be of brass
- motto etched centrally on the blade – on original examples the motto was sited much closer to the hilt than to the tip, a characteristic common to all Third Reich daggers with blade mottoes
- a maker's mark other than that of Carl Eickhorn, who was the sole producer of the SA Feldherrnhalle dagger
- blades etched with a spurious dedication from Viktor Lutze – not known on originals

## THE SS 1933 DAGGER

Reproductions have features similar to those of fakes of the SA 1933 dagger.

**183**. *A poorly fitting blade is clearly evident on this reproduction* SS *dagger.*

## THE SS 1934 RÖHM HONOUR DAGGER

As for the *SA* 1934 Röhm Honour Dagger.

## THE SS 1934 HIMMLER HONOUR DAGGER

As for the *SS* 1933 dagger, but with these additional features:

- poorly etched Himmler dedications, sometimes even without the top half of the 'T' in 'Himmler' – presentation etching was always strictly controlled and of excellent quality
- engraved Himmler dedications – originals were never engraved
- offset dedications – the dedication should be central along the blade's reverse spine
- *RZM*-marked blades – unknown on original Himmler Honour Daggers
- blades bearing non-Eickhorn trademarks – Carl Eickhorn had sole manufacturing rights to the Himmler Honour Dagger. Only 200 were made and each was serially numbered

## THE SS 1936 CHAINED DAGGER

As for the *SS* 1933 dagger, but with these extra features:

- poorly detailed or ill-fitting central scabbard mounts – authentic mounts were sharply die-struck with high-relief swastikas and background stippling
- roughly cast chain links bearing pock marks and other flaws – originals were crisply die-stamped in nickel silver or nickel-plated steel
- inspection stamps to the reverse lower crossguard, the sign of a 'parts' assembly – crossguard stamps were applied only to pre-*RZM SS* 1933-pattern daggers.
- makers' marks – the *SS* 1936 chained dagger was not maker marked.

## THE SS 1936 HONOUR DAGGER

As for the *SA* 1935 Honour Dagger, and again Eickhorn was the only producer of this pattern. None should have 'Himmler' blades. In addition, beware of exquisitely hand-made items with ivory grips allegedly presented to 'Sepp' Dietrich, Julius Schreck and other *SS* leaders. A few of these pieces have the would-be recipient's initials at the top of the grip, instead of the *SS* runes. All are postwar creations.

## THE NSKK 1933 DAGGER

Reproductions have features similar to those of fakes of the *SA* 1933 dagger.

## THE NSKK 1936 CHAINED DAGGER

As for the *NSKK* 1933 dagger, but with the following additional indicators:

- loosely attached scabbard central mounts – genuine mounts were made to fit well, with no large gaps
- very thin raised borders to the central mount – original lips were usually quite substantial and in any case always matched those on the corresponding locket and chape; a mismatch may mean that the piece under consideration is simply a 1933 dagger with spurious postwar chains and accompanying central mount added
- poorly defined suspension chain links, cast in a soft alloy – originals were well struck in either nickel silver or nickel-plated steel
- circular junction holes to the chain links – authentic examples were domed with a flat base

- chain links bearing the spurious inscription *Muster geschützt NSKK Korpsführer* ('Pattern copyright of the *NSKK* Supreme Commander') – original chains were stamped *Musterschutz NSKK-Korpsführung* ('Copyright of the *NSKK* High Command') on the reverse of one of the two upper links only; the back of the chain's other upper link usually featured the *RZM* code number 'M5/8' of the chain manufacturer, Assmann.
- a maker's mark other than that of Carl Eickhorn, sole authorised producer of the *NSKK* 1936 chained dagger.

## THE *NPEA* 1935 STUDENT'S AND STAFF DAGGERS

- plastic grips – originals were carved from a variety of woods
- badly etched, oversized or disjointed blade mottoes – genuine etching was always neat (smaller than that of the *SA*, *SS* and *NSKK* mottoes), uniform and of high quality
- a depressed or 'scooped out' background to the blade motto, indicative of an original *SA* blade which has had its motto erased and replaced by that of the *NPEA*.
- crossguards bearing *SA* inspection marks – again commonly encountered on so-called *NPEA* daggers which are, in fact, original *SA* daggers with re-etched blades and other minor modifications
- blades bearing a trademark other than those of Karl Burgsmüller, Carl Eickhorn or Max Weyersberg – further evidence of conversion, since Burgsmüller was the sole distributor, and Eickhorn and Weyersberg the only manufacturers, of *NPEA* daggers
- separate scabbard tips soldered into position – yet another sign of the converted *SA* dagger; original *NPEA* 1935 scabbards were formed in a single piece, with an integral hollow tip
- 'Germany' stamped on the blade ricasso or tang – the sign of a postwar copy made for export
- sharp blade spines – genuine *NPEA* blades had rounded spines
- scabbard liners in a thin grey alloy – originals were wide and made of brass

- four grooves to the scabbard's frog stud – originals had only two grooves

## THE *NPEA* 1936 STAFF CHAINED DAGGER

As for the *NPEA* 1935 student's and staff daggers (except the sixth and tenth points).

## THE HITLER YOUTH 1933 KNIFE

- grip insignia glued into position – originals had the Hitler Youth diamond secured by means of two prongs which were pushed through retaining slots in the obverse grip facing and then bent over, prior to assembly of the handle
- blades stamped with the word 'Solingen' on its own – a feature not encountered on Third Reich pieces but commonly found on modern Scout knives, many of which are identical in pattern to the Hitler Youth 1933 knife, and are easily converted by removing the fleur-de-lis diamond from the grip and its replacement by fake Nazi grip insignia
- post-1938 dated blades etched with the Hitler Youth motto – the motto appeared only on knives produced prior to August 1938
- alloy scabbards – these should be made from steel
- scabbards with a raised lip at the throat – originals had no lip

## THE HITLER YOUTH LEADER'S 1937 DAGGER

- badly cast alloy pommels – originals were made from nickel-plated steel or aluminium, with crisp high-relief Hitler Youth insignia
- plastic grips – each genuine handle took the form of a carved wood base wrapped overall in fine silver wire
- blades bearing the spurious etched dedication *Für verdienste um die Deutsche Jugend* ('For merit in the German youth movement') – not known on originals
- cheap-quality scabbard coverings – genuine scabbards were finished in high-grade leather
- poorly detailed locket eagles – these should be well executed

- locket eagles holding swords and spanners – originals held swords and hammers
- locket eagles in a stationary pose – genuine examples were depicted 'taking off' towards the viewer's right

## THE *DLV* 1934 KNIFE AND THE *NSFK* 1937 KNIFE

- roughly cast fittings – originals were devoid of flaws
- painted swastikas to the crossguard – these should be finely executed in black enamel
- unmarked scabbard lockets – genuine lockets bore the *DLV* stamp or the *NSFK* stamp, or both
- scabbard lockets with attachment screws at the lower end – originals had these screws situated at the top, immediately beneath the throat
- scabbard chapes with spherical tips – originals were flattened to produce an oval effect

## THE DIPLOMATIC CORPS 1938 DAGGER AND THE GOVERNMENT ADMINISTRATION 1939 DAGGER

- lightweight alloy fittings – original parts were in silver-plated brass
- unmarked fittings – each genuine component was stamped during manufacture with an identification number, positioned so as to be hidden from view when the dagger was assembled
- white plastic grips – these should be genuine or simulated mother-of-pearl over wood
- crossguard reverses which are rough and unplated, or which feature two circular ejector pin marks – original crossguards had smooth plated reverses
- rounded edges to the flats of the blade – these should be well defined
- *RZM*-marked blades – neither Diplomatic Corps nor government administration daggers were *RZM* marked
- scabbards with separate tips and a distinctive swirling pattern to the pebbling – genuine scabbards were constructed in one piece, with dense and irregular pebbling
- scabbard throats soldered into place – originals were affixed by means of two screws

## THE WATERWAYS PROTECTION POLICE 1938 DAGGER

The Waterways Protection Police dagger was based largely on that of the Navy, and fakes of it parallel those of the latter, as already described.

## THE FIRE BRIGADE 1870 DAGGER AND THE FIRE BRIGADE 1870 AXE

Both of these items were introduced prior to the Nazi period, and their original designs remained unchanged during the Third Reich. They continue to be manufactured, in identical styles, for wear by today's firemen on formal occasions. Short of presentation inscriptions, or dated marks, there is nothing to distinguish a Nazi Fire Brigade dagger or axe from an imperial or post-1945 one.

## THE LAND CUSTOMS 1937 DAGGER

- heavy alloy or brass hilt fittings – originals were in nickel-plated steel or aluminium
- grip grooves and wire running in the wrong direction, i.e. from high left to low right – they should run from high right to low left
- grip wire in the form of a twisted double strand, bordered on either side by a single strand – the original style was simply a twisted double strand
- rounded edges to the flats of the blade – these should be sharply defined
- *RZM* marks on the blade – Customs daggers were never *RZM* marked
- blades bearing a lightly etched version of the F.W. Höller trademark – this is the most common mark found on reproduction Customs daggers, but original Höller logos were deeply etched or stamped, with fine detailing to the thermometer
- plastic scabbards – genuine examples were leather over steel
- narrow brass liners soldered into the scabbard – these should be of steel or zinc alloy, riveted into position

## THE WATER CUSTOMS 1937 DAGGER

As for the Land Customs 1937 Dagger, but with gold-plated bronze fittings.

## THE POSTAL PROTECTION CORPS 1939 DAGGER

- soft wood handles coated with black ink which can easily be washed off – original grips were in hard wood, permanently stained black
- poor detailing to the grip insignia – such badges were always well formed
- crossguards featuring painted swastikas – originals were enamelled
- hilt and scabbard fittings roughly cast in brass with a light silver plating – genuine examples were finely executed in nickel silver or nickel-plated steel
- blades bearing a maker's mark other than that of Paul Weyersberg – Weyersberg was the sole producer of the Postschutz dagger

## THE RAILWAY GUARD 1938 DAGGER

- brass hilt fittings – never seen on authentic Railway Guard daggers; genuine fittings were die-cast in aluminium
- rounded edges to the flats of the blade – these should be sharply defined
- heavy, grey-coloured steel scabbards – originals were made from aluminium and were always lightweight
- loosely attached scabbard liners in wood or plastic – these should be of steel or zinc alloy and securely riveted into position

## THE *RAD* 1934 HEWER

- crudely cast, chrome-plated alloy hilts – originals were expertly cast in nickel silver or nickel-plated steel
- Roughly etched blade mottoes, with a stippled background to the lettering – etching should be smooth and well defined
- mottoes etched so as to read from the tip of the

blade towards the ricasso – genuine *RAD* mottoes were etched to be read from the hilt end of the blade towards the tip (i.e. unlike those of the *SA*, *SS*, *NSKK* and *NPEA*)
- *RZM*-marked blades – *RAD* sidearms were never *RZM* stamped
- ill-formed designs to the scabbard mounts – these should be well defined
- lockets and chapes in a lightweight grey alloy – originals were in nickel silver or nickel-plated steel

## THE *RAD* LEADER'S 1937 HEWER

- hilts cast in a heavy grey alloy – originals were in either nickel-plated steel or aluminium
- as the second to fifth points for the *RAD* 1934 Hewer

## THE *RLB* 1936 KNIFE, THE *RLB* 1936 DAGGER, THE *RLB* 1938 KNIFE AND THE *RLB* 1938 DAGGER

As for the Postal Protection Corps 1939 Dagger, points 1, 2 and 4.

## THE *TENO* 1938 HEWER

- pitted hilts, crudely cast in chrome-plated brass or alloy – originals were finely crafted from nickel-plated steel
- unmarked grips – each original grip plate had the Eickhorn trademark stamped on its reverse prior to assembly
- blades featuring a maker's mark other than that of Carl Eickhorn, the patentee of all *Teno* edged weapons
- poorly etched blade markings – authentic representations of the Eickhorn logo and *Teno* eagle were well executed
- *RZM*-marked blades – *Teno* blades were never *RZM* marked
- thin scabbard liners soldered to the inside of the locket – genuine liners were riveted into position

### The *Teno* 1938 Dagger

As for the *Teno* 1938 Hewer, the third to sixth points, with these additional indicators:

- badly formed hilt fittings in a variety of plated alloys – originals were keenly cast in aluminium, oxidised to give a tarnished appearance
- scabbards in silver-plated brass – genuine examples were in aluminium or oxidised steel
- scabbards with elongated tips – the chape should terminate in a flattened oval
- blades bearing etched dedications from Dr Fritz Todt or *SS-Gruppenführer* Hans Weinreich – never featured on originals

### The Red Cross 1938 Hewer

- poorly cast aluminium hilt fittings – originals were in nickel silver or nickel-plated steel
- wooden grips – genuine examples were in black bakelite

### The Red Cross 1938 Dagger

- lack of detail to the crossguard cartouche – originals were crisply defined
- *RZM*-marked blades – Red Cross sidearms were never *RZM* marked

### The Forestry Service 1938 Cutlass

- hilt fittings in glittery gilt alloys – fittings made between 1938 and 1942 were in gold-plated brass, bronze or aluminium
- plastic handles – originals were in staghorn, ivory or white celluloid
- plastic scabbards – these should be leather

### The Hunting Association 1936 Cutlass

- hilt fittings in glittery silver alloys – fittings made between 1936 and 1942 were in silver-plated brass, bronze or aluminium
- plastic handles – originals were in staghorn or wood
- plastic scabbards – these should be leather

### The Rifle Association 1939 Cutlass

- ill-formed grip rifles – originals were expertly shaped
- *D.Sch.V.* insignia attached to the shell guard by two long pins – originals were fixed by two short pins or a single rivet

In the field of daggers, then, many anomalies exist, and differentiating between original items and reproductions can be a perplexing business. Certain blades, for example those of the Army and Railway Guard, should have sharp ridges, while with others, such as those of the *SA*, *SS* and *NSKK*, a sharp ridge is indicative of faking. Genuine Diplomatic Corps fittings were manufactured in silver-plated brass, but *Postschutz* and *Luftschutz* daggers with components in silver-plated brass are likely to be counterfeit. Red Cross hewer grips should be in plastic, not wood, whereas genuine hunting sidearms had grips in wood, not plastic. Original first-pattern *Luftwaffe* daggers never had grip wire of a single twisted double strand, but Customs Service daggers always did. And so it goes on. Fortunately, however, there are other more general factors to be considered when examining pieces. These are outlined below.

### Assembly

Original dagger assembly techniques differed from one pattern to another, but in all cases components fitted well together, as quality-control in German arms factories was meticulous. Modern reproductions are no match for originals in this respect and so collectors must be wary of daggers with any misfitting parts. Blades should never be loose, and authentic pieces with large gaps between the pommel and the grip, or between the grip and the crossguard, would have been scrapped at once. Likewise, an *SS* grip with runes and eagle insignia lying in hollows far too large or too shallow to accommodate them would never have been passed by the *RZM* inspectors. The fundamental rule of

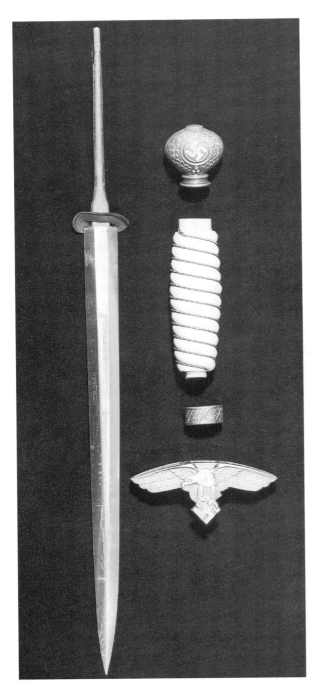

**184**. *Disassembled* Luftwaffe *1937 dagger, showing the main component parts of blade, pommel, grip, ferrule and crossguard.*

**185**. *Two blade tangs. The squared-off example on the left is a reproduction, while that on the right, showing the billet clamp seam running along its rounder edge, is original.*

thumb is that if a dagger lacks refinement to such an extent that it would have shamed a Nazi uniform, then it is most probably a postwar copy.

Collectors should not be reluctant to strip pieces down prior to purchase, because internal examination plays a crucial part in fake detection. It is only by disassembly that the presence or otherwise of

Diplomatic Corps and Government Administration numbers or of Eickhorn trademarks on the grips of *Teno* hewers can be determined. Above all, inspection of the blade tang is most important, since fake blades tend to be without the billet clamp seam, a ridge which ran along the sides of the vast majority of original tangs. Genuine Navy tangs were ground down to facilitate attachment of the cross-guard retaining spring, however, so did not have the billet clamp seam. The same was true of Waterways Protection Police daggers, Rifle Association cutlasses and most swords, which required slightly curved tangs and so had their seams ground off during assembly.

### MARKINGS

The presence of a known maker's trademark or other stamping should never be viewed as a guarantee of authenticity, as every form of mark has been faked. Moreover, various spurious company logos that never existed at all during the Third Reich have been created in recent years, including an 'MH' monogram which features on many fake Army, *Luftwaffe* 1937-pattern, Diplomatic Corps and government administration daggers. As a general maxim, fake marks tend to be more 'sketchy' in appearance than originals, which were always well defined. With the notable exception of the 'Karl Burgsmüller' *NPEA* daggers, any daggers purporting to emanate from Berlin should be viewed with particular suspicion, since 'Berlin' frequently replaces 'Solingen' on reproduction trademarks as a convenient way of circumventing the copyright restrictions protecting their genuine counterparts.

### ENGRAVING AND ETCHING

Fakers have for long 'enhanced' original Nazi daggers by the application of elaborate postwar designs and dedications, which may be either engraved or etched. Pieces so altered are then offered as unique Honour Daggers or presentation blades, and can fetch two or three times the price sought for a basic, unembellished dagger. Once again, the major factor to consider when

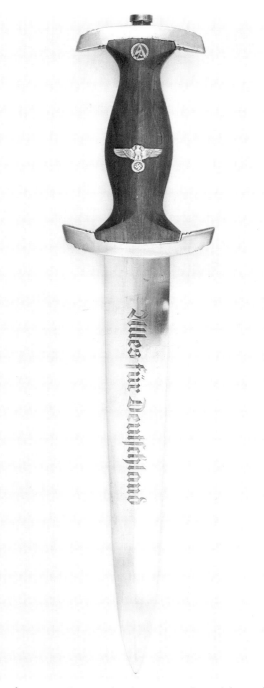

**186**. *A typical example of a postwar 'parts' dagger. The hilt is original but the blade is a reproduction, with the motto facing the wrong way.*

173

examining ornamentation like this is quality. Most reproductions simply fail to come up to the standard of workmanship required during the Third Reich. Complete etched blades have also been faked.

Originally, engraving tended to be confined to 'one-off' presentation pieces, although it was also favoured by many dagger owners who had their names or initials applied to the crossguard reverse for identification purposes. Inscriptions such as these were executed by local jewellers, by hand. Any engraving, therefore, which is not professional in appearance, which has been done mechanically, or which has not aged with the rest of the dagger, is likely to be a modern addition.

Etching, on the other hand, was a standard feature of many Nazi daggers and a wide range of genuine patterns and finishes was produced. Some etchings had light grey backgrounds to their designs, while others had recesses which were almost black. The etched mottoes on *SA*, *SS*, *NSKK* and *NPEA* daggers were consistently sited much closer to the hilt than to the tip of the blade, whereas fake mottoes are often centred. Unfortunately, reproductions of every type of etching exist. The following selection gives an idea of the varieties of copy which may be encountered:

- fake Army 1935 dagger blades etched with eagles and foliage on the obverse, dedications such as *Gestiftet von der Kameradschaft 155er zu Breslau* or *Für Verdienste im Generalstab* on the reverse, and the Alcoso trademark
- fake *Luftwaffe* 1937 dagger blades with eagles and foliage on the obverse, dedications such as *Ehrenpreisschiessen der Luftwaffen-Gruppe 3* or *Geschwader Mölders* on the reverse, and Alcoso or Eickhorn trademarks
- fake *SA* 1933 dagger blades with the trademarks of Ernst Pack, Paul Weyersberg or '*RZM M7/12*'
- fake *SA* and *SS* 1934 Röhm Honour Dagger blades with Eickhorn marks
- fake *SA* and *SS* 1934 Röhm Honour Dagger blades in artificial Damascus steel with raised gilded lettering, the Lüneschloss trademark and the *Echt Damast* ('Genuine Damascus') designation

- fake *SS* 1933 dagger blades marked '*RZM 1211/39 SS*' or '*RZM 941/40 SS*'
- fake *NPEA* dagger blades with the Burgsmüller mark

All of the above blades have been made to a relatively high standard, but many others circulate in a range of different qualities. Points to be especially wary of are:

- incomplete, deformed or uneven etching – all patterns should be crisp, symmetrical and flawless. Lettering with chunks missing, or with ragged outlines, would never have been allowed to leave any Nazi blade factory
- stippling, or small raised dots, either on the etching itself or on the background area – genuine etching was always smooth
- painted lettering – fake *SA* and *RAD* sidearms in particular have been encountered with their etched mottoes painted brown as a means of masking the stippling described above; paint was never applied to original blades. Many copies bear the trademark of Gebrüder Heller, Schmalkalden.

### 'FANTASY' DAGGERS

The lowest of the low, so far as purists are concerned, are the so-called 'fantasy' daggers, which cannot be classed even as reproductions since they had no original counterparts. These fantasies comprise various combinations of reworked dagger parts, both genuine and fake, Nazi and otherwise, and are in themselves visually attractive. However, all are readily identifiable as postwar creations by their distinctive and undocumented design characteristics.

# SWORDS

Nazi swords have not been reproduced to anything like the same extent as daggers, partly because they are more expensive to manufacture and partly because they are generally less popular with collectors. A number of original *Wehrmacht* and other common pieces have had spurious postwar dedications etched on their blades in an effort to upgrade them and make them more saleable, but the poor quality of workmanship involved usually reveals such alterations for what they really are. The following paragraphs describe fakes which are known to have been made of the scarcer patterns of sword.

### THE *LUFTWAFFE* GENERAL'S SWORD

A few genuine examples of the standard *Luftwaffe* general's 1935- and 1937-pattern swords, which are rarities in their own right, had their blades etched during the 1970s with the relevant Göring presentation dedication to 'convert' them into Swords of Honour. Whereas the authentic gilded etching was distinguished by its blue background following the contours of the bordering white nickel scrollwork, the postwar version had the background terminating incorrectly as a straight line across the blade.

### THE *FELDJÄGERKORPS* 1933 SABRE

The original *SA-Feldjägerkorps Säbel* had a plain blade. Collectors should be wary of any swords etched with the *SA* motto *Alles für Deutschland*, which is indicative of a postwar Army conversion.

### SS SWORDS

Of all Third Reich sword patterns, those of the *SS* have long been in greatest demand and shortest supply. It is not surprising, therefore, that they have received most attention from fakers over the years, particularly since readily obtainable original Army sabres and Police *Degen* can be converted easily into *SS* items. Spurious types include:

- basic army ordnance sabres with *SS* runes etched on to the obverse langet
- silver-plated lion-head sabres (an authorised Army option) with enamelled *SS* runes insignia soldered to the reverse langet

- sabres as above which have had the blade etched with the *SS* motto *Meine Ehre heisst Treue*
- standard Police officers' 1936 swords with the grip eagle replaced by *SS* runes. Original *SS* officers' 1936-pattern swords were either unmarked or bore the logo of Peter Daniel Krebs. So-called *SS* 1936 sword blades featuring the trademarks of Alcoso, Carl Eickhorn, Robert Klaas, Carl Julius Krebs, Ernst Pack, Hermann Rath, Emil Voos, WKC or any of the many other Police sword makers are typical of such conversions.
- converted items as above with the blade incorrectly etched with the *SS* motto or meaningless inscriptions such as *SS Sicherheitshauptamt* or *In Kameradschaft, H. Himmler, Reichsführer-SS*

Complete reproductions of the *SS* officer's 1936 sword appeared on the market during the 1980s, but are very poor in quality and finish when compared to original or converted Police items. Such copies usually bear the maker's mark of F.W. Höller.

It is also noteworthy that the former *SS* Damascus swordsmith Paul Müller was very active during the 1960s in the creation of numerous reproduction Nazi Damascus blades for bogus *SS* Birthday Swords and Honour Daggers. His fakes usually bore the raised and gilded inscription *Echt Damast, P. Müller*. He continued to practise his craft until six months before his death in 1971, at the age of 87.

### JUSTICE AND PRISON SERVICE SABRES

Fakes of the Justice and Prison Service 1936 sabres were produced during the 1970s by marrying crudely cast reproduction brass hilts with standard original Army *Säbel* blades, grips and scabbards. Most genuine Justice and Prison sabres were made by Clemen & Jung, Höller, Eickhorn and WKC, so other trademarks are signs to be cautious of.

These examples are not exhaustive, but serve to provide the reader with an indication of the range of spurious swords currently on the market.

# BAYONETS

Bayonets have not been faked to any great degree but, like swords, are particularly susceptible to the postwar addition of distinctive grip insignia. Many standard Army M.98 *Extra-Seitengewehre*, for example, have been easily converted in this way to pass for rare *HJ* or *RLB* bayonets. In such cases, the offending bogus emblems tend to be glued into place rather than being attached by bent pins as originals were, since the affixing of badges in the correct manner would necessitate carefully removing the bakelite grip plates and re-riveting them into position.

Various completely fanciful etchings and engravings have also appeared on original *Wehrmacht* dress bayonets and even M.84/98 service bayonets over the years. The inscriptions involved are generally quite meaningless: 'SS Adolf Hitler', 'Reichsmarschall', 'Wallenstein', 'Siegfried' and 'SS Totenkopfwachsturmbann Sobibor' are examples. Covered with over-sized Sig-runes and morale-boosting slogans, these decorated bayonets are base in the extreme and are instantly recognisable by virtue of their inherent vulgarity.

# KNIVES

Some fighting knives were similarly 'upgraded' after 1945 by the addition of fake insignia, etching and engraving. Once again, *SS*-related material, including death's heads and runic badges, are usually involved. Other reproductions such as the *Reichskanzler Adolf Hitler* and *Deutschland Erwacht* penknives, which were struck in the 1970s from previously unused dies produced in 1933 by the J.A. Henckels firm, also have somehistorical basis. The majority of fake Nazi knives, however, are made from scratch, the raw results of over-active post-1945 imaginations. The 'SS-treifendienst Pantograph knife', for instance, and the 'Belt Buckle Dagger', examples of which suddenly flooded the market in the 1970s, represent all that is bad in the fantasy. Crude in design, construction and quality, it is inconceivable that anything remotely like them would have seen the light of day in Hitler's Germany. In a similar vein, modern sheath knives with cast copper eagles on the grip and etched swastikas on the blade have been offered for sale as 'rare prototypes' and 'important finds'. Crude fakes like these should always be avoided unless the enthusiast wishes to utterly debase his collection.

# ACCOUTREMENTS AND ACCESSORIES

As well as the edged weapons themselves, all types of accoutrements and accessories have been faked.

**187.** *The so-called* Luftwaffe *'Belt Buckle Dagger', produced in substantial quantities during the early 1970s. Similar pieces were attributed to the Army, Navy, Waffen-SS and even the Red Cross!*

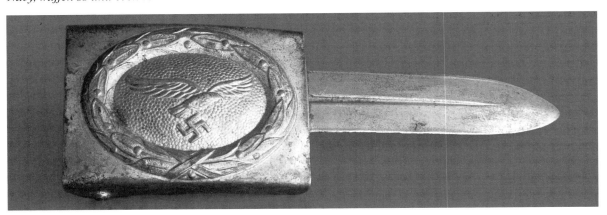

**188**. *The reverse of the 'Belt Buckle Dagger', revealing a spurious stamped Eickhorn trademark. The fact that such items were manufactured from rusted steel gave them an aged appearance even when new.*

Rare and expensive hangings straps such as those for the *Bahnschutz* and *Teno* daggers, and even common Army and *Luftwaffe* hangers, have been widely copied, often with fittings in a very lightweight alloy exhibiting a plastic-like appearance. Such reproductions exist in a range of qualities, some utilising leftover stocks of genuine components while others feature a combination of original and fake parts or fittings from official post-1945 East German daggers. Signs to be particularly wary of are obvious mismatching, very new leatherwork and East German inked marks including 5- or 6-digit unit or production numbers.

Fake dagger knots in aluminium wire are extremely difficult to detect, particularly since wrapping them in rubber bands for twenty-four hours can endow them with a patina giving the impression of great age. However, copies typically have slightly narrower shanks than originals, with a more open lattice weave, and the top part of the acorn is usually not as neatly finished as it should be. *SS* sword knots have been extensively copied as well, with reproductions featuring poorly formed runes on the officers' and junior NCOs' types, and silver wire borders instead of the correct double white edge stripes on the enlisted ranks' version.

Factory guarantee tags, storage bags and miniature weapons have all received attention from the fakers, and imaginative 'presentation cases' embellished with eagles, swastikas, *SS* runes, Röhm dedications and so on created from scratch to enhance appropriate daggers and knives. Even Nazi cutlers' sales catalogues have been reprinted in recent years for the benefit of collectors and researchers.

Contrary to popular belief, not all antique dealers, or indeed militaria dealers, are experts in their field. It is true to say that they, like collectors, do get caught out by fakes from time to time. However, most dealers are honest, and when asked if a piece is authentic or not they are obliged to make their position clear. There is a world of difference between a dealer who deliberately defrauds the public and one who merely errs in judgement concerning authenticity. Even if he is genuinely uncertain, the dealer can express an opinion. At this point, of course, the ultimate onus of satisfaction returns to the buyer, and even among the experts there will invariably be areas of disagreement or contention. In short, money-back guarantees mean little in these circumstances and '*Caveat emptor*' still holds good. If the would-be buyer has any 'gut feeling' reservations about the originality of a piece offered for sale he should leave it alone, for an honest collector will readily admit that the vast majority of his 'maybes' turned out to be fakes. If a piece looks as if it could have been made yesterday, it probably was!

# BIBLIOGRAPHY

The following books are recommended further reading on the subject of Third Reich edged weapons and the organisations which wore them.

Angolia, J.R., *Swords of Germany, 1900-1945*, Bender, San José, 1988

Angolia, J.R. and Littlejohn, D., *NSKK & NSFK*, Bender, San José, 1994

Bowman, J.A., *Third Reich Daggers, 1933-1945*, Imperial Publications, Lancaster, 1990

Bowman, J.A., *Third Reich Dagger Reproductions*, Imperial Publications, Lancaster, 1993

Catella, F., *Waffenfabrik Katalog*, published privately, France, 1986

Davis, B.L., *German Uniforms of the Third Reich, 1933-1945*, Blandford, Poole, 1980

Halcomb, J., *The SA*, Agincourt, Columbia, 1985

Halcomb, J., *Uniforms and Insignia of the German Foreign Office and Government Ministries, 1938-1945*, Agincourt, Columbia, 1984

Hughes, G.A., *German Military Fighting Knives, 1914-1945*, Imperial Publications, Lancaster, 1992

Johnson, T.M. et al., *Collecting the Edged Weapons of the Third Reich, Vols. 1-8*, published privately, Columbia, 1975-98

Johnson, T.M., *Wearing the Edged Weapons of the Third Reich*, published privately, Columbia, 1979

Johnson, T.M. and Bradach, W., *Third Reich Edged Weapon Accoutrements*, published privately, Columbia, 1978

Klietmann, Dr K.G., *German Daggers and Dress Sidearms of World War II*, Field & Fireside, Virginia, 1967

Littlejohn, D., *The Hitler Youth*, Agincourt, Columbia, 1988

Lumsden, R., *Third Reich Militaria*, Ian Allan, Shepperton, 1987

Lumsden, R., *Detecting the Fakes*, Ian Allan, Shepperton, 1989

Lumsden, R., *The Black Corps*, Ian Allan, Shepperton, 1992

Lumsden, R., *The Waffen-SS*, Ian Allan, Shepperton, 1994

Lumsden, R., *The Allgemeine-SS*, Reed/Osprey, London, 1993

Lumsden, R., *SS Regalia*, Bison, London, 1995

Lumsden, R., *Himmler's Black Order*, Sutton, Stroud, 1997

Mermet, C. and Marfault, J., *Les Dagues du III$^e$ Reich*, Éditions du Portail, La Tour, 1981

Mollo, A., *Daggers of the Third German Reich, 1933-1945*, Historical Research Unit, London, 1967

Mollo, A., *German Uniforms of World War 2*, MacDonald & Jane's, London, 1976

Stephens, F.J., *Daggers, Swords & Bayonets of the Third Reich*, Patrick Stephens, London, 1989

Stephens, F.J., *Reproduction? Recognition!*, Military Collector Inc., Sepulveda, 1981

Venner, D. and Tavard, C., *Les Armes Blanches du III$^e$ Reich*, Grancher, Paris, 1977

Walker, G.L. and Weinand, R.J., *German Clamshells and Other Bayonets*, published privately, Quincy, 1985

Weinand, R.J., *NPEA Daggers*, published privately, Quincy, 1988

# INDEX